Aristotle's Ethics

Continuum Studies in Ancient Philosophy
Series Editor: James Fieser, University of Tennessee at Martin, USA

Continuum Studies in Ancient Philosophy is a major monograph series from Continuum. The series features first-class scholarly research monographs across the field of Ancient Philosophy. Each work makes a major contribution to the field of philosophical research.

Aristotle and Rational Discovery, Russell Winslow
Aristotle's Metaphysics, Jeremy Kirby
Aristotle's Theory of Knowledge, Thomas Kiefer
The Enduring Significance of Parmenides, Raymond Tallis
Happiness and Greek Ethical Thought, M. Andrew Holowchak
The Ideas of Socrates, Matthew S. Linck
Parmenides and To Eon, Lisa Atwood Wilkinson
Plato, Metaphysics and the Forms, Francis A. Grabowski III
Plato's Stepping Stones, Michael Cormack
Pleasure in Aristotle's Ethics, Michael Weinman
Pythagoras and the Doctrine of Transmigration, James Luchte
The Socratic Method, Rebecca Bensen Cain
Stoic Ethics, William O. Stephens
Stoic Virtues, Christoph Jedan

Aristotle's Ethics

Moral Development and Human Nature

Hope May

continuum

Continuum International Publishing Group
The Tower Building, 11 York Road, London SE1 7NX
80 Maiden Lane, Suite 704, New York NY 10038

www.continuumbooks.com

British Library Cataloguing-in-Publication Data
A catalogue record for this book is available from the British Library.

ISBN: HB: 978-0-8264-9110-7

Library of Congress Cataloging-in-Publication Data
May, Hope.
Aristotle's Ethics: moral development and human nature/Hope May.
p. cm.
Includes bibliographical references (p.) and index.
ISBN 978-0-8264-9110-7
1. Aristotle. Nicomachean ethics. 2. Moral development. 3. Happiness. I. Title.

B430.M39 2009
171'.3–dc22 2009016262

Typeset by Newgen Imaging Systems Pvt Ltd, Chennai, India
Printed and bound in Great Britain by the MPG Books Group

To Jeffrey Wigand

Beloved husband, esteemed mentor, cherished friend

We are masters of our actions from the beginning right to the end.

Aristotle (Nicomachean Ethics 1114b*)*

He who lets the world, or his own portion of it, choose his plan of life for him has no need of any other faculty than the ape-like one of imitation. He who chooses his plan for himself employs all of his faculties. He must use observation to see, reasoning and judgment to foresee, activity to gather materials for decision, discrimination to decide, and when he has decided, firmness and self-control to hold to his deliberate decision.

John Stuart Mill, 1859

To guide our own craft, we must be captain, pilot, engineer; with chart and compass to stand at the wheel; to match the wind and waves and know when to take in the sail, and to read the signs in the firmament over all. It matters not whether the solitary voyager is man or woman.

Elizabeth Cady Stanton, 1892

Creating anything is hard. It's a cliché thing to say, but every time you start a job, you just don't know anything. I mean, I can break something down, but ultimately I don't know anything when I start . . . You start stabbing out, and you make a mistake, and it's not right, and then you try again and again. The key is you have to commit. And that's hard because you have to find what it is you are committing to.

Philip Seymour Hoffman, 2008

Contents

Preface and Acknowledgments xi
Abbreviations and Notes on Translations xiv

1 The Intellectualism Debate 1
- Introduction 1
- The Function Argument 5
  - Goods and goals 6
  - The human good (i.e., goal) 7
- The Difficulty of Book X 11
  - Intellectualism 14
  - Inclusivism 15
  - Developmentalism 17

2 A Developmentalist Interpretation of the Function Argument 18
- Introduction 18
- Causality in Man-made Production 19
  - Hypothetical necessity in man-made production 21
- The Craftsman's Motivational System 23
  - The intellectual components 23
  - Strategic planning 23
  - Form or knowledge 24
  - The desiderative components 26
  - Choice 27
  - Wish 28
  - Wish and pleasure 29
  - Calling 30
  - Calling and pleasure 30
- Natural Production 33
  - Calling vs target 33
  - The role of soul 35
  - Soul as form 36
- The Function Argument Revisited 38
- The Human Being's Motivational System 40

3 Ethical Virtue as a Developmental Prerequisite for Contemplative Excellence 46
Introduction 46
The Nature and Development of Affective Virtue 48
The uniqueness of man's desiderative faculty 49
The uniqueness of man's actions 50
Affect and the desiderative components 52
Affect as an efficient cause of action 53
Affect as a final cause of action 53
Affect as the most final cause of action 54
The primacy of affect 55
The particularity of affect 56
Correct affect 57
Affect and the target of nature 59
The incorrectness of man's default affective state 60
Habituation: correcting man's default affective state 60
The public component in affective training 61
The requirement of pleasure 63
The private component in affective training 64
The awakening of proper pleasure 66
The role of law 69
The developmental role of proper pleasure 70
The Development of Contemplative Excellence 70
The motivational preconditions of contemplative excellence 71
The cognitive preconditions of contemplative excellence 72
The origin of universals 73
The neurobiology of memory 74
The activities on which knowledge depends 76
Securing the Development of Contemplative Excellence 79
Phronêsis 80
Ethical virtue: nourishing and safeguarding contemplative excellence 83
The intellectualism debate revisited 84
The Development of the Students 85
The transformative aim of the *Nicomachean Ethics* 85
Conclusion 87

4 Aristotle's Developmentalist Approach Modernized: Flourishing as Self-Concordance 90
Is Aristotle's Theory of Flourishing Still Relevant? 90
Flourishing and autonomy: the modern view 91
Virtue Theory 93
Virtuous self-determination 95
Positive Psychology 96
The hedonic approach to flourishing 97
The *eudaimonic* approach to flourishing 98
Self-Determination Theory 98

Needs and desires 99
The need for autonomy 100
Varieties of teleological experience 101
Intrinsic and extrinsic motivation 102
Self-concordant activity 102
The self-perception of autonomy 103
The pleasures of self-concordance 104
Self-Determination Theory and Aristotle's Ethics 104
Aristotle on intrinsic motivation 104
Aristotle on the pleasure of autonomy 107
The ownership of one's virtue 111
The effects of reflecting on one's virtue 112
Autonomy and habituation 113
Primitive and normative habits 113
Internalization and integration of norms 114

5 The Developmental Preconditions of Self-Concordance 119
Introduction 119
Executive Function 120
The Intellectual Component of Executive Function 121
The Navigational Component of Executive Function 122
Self-regulation 123
Private speech 123
The development of private speech 124
The steering function of private speech 125
Aristotle on private speech 126
Self-regulation and habit 128
The ubiquity of automaticity 128
Masterful self-regulation: mindfulness of habit triggers 130
Masterful self-regulation: classification of habit triggers 130
Masterful self-regulation: steering attention 131
Masterful self-regulation: steering attention inwards 133
Masterful self-regulation: forming implementation intentions 136
Masterful self-regulation and self-concordance 140
The Motivational Component of Executive Function 141
Framing 142
Hope and self-efficacy 143
Resilience 143
Masterful motivation: using private speech to motivate 144
Aristotle's Approach Modernized 147
The modern virtues 147
The awakening of proper pleasure 149

6 The Ecological Preconditions of Self-Concordance 153
Introduction 153
The role of legal ecology 154
Programs that cultivate 157

Conclusion 161
Wendy Kopp: An Illustration of Self-Concordance 163

Notes 166
Bibliography 174
Index of Names 181
Index of Subjects 182

Preface and Acknowledgments

This book has two primary aims. The first is to provide an interpretation of Aristotle's *Nicomachean Ethics* (*NE*) which is grounded in his views about the biological development of human beings. There is notorious scholarly debate about Aristotle's conception of *eudaimonia*. While it is clear that contemplative activity plays a fundamental role in Aristotle's theory of flourishing, some scholars contentiously claim it is the *only* fundamental component. On this view known as "intellectualism," morally virtuous activity is *not* a fundamental component of the flourishing life—a conclusion which many scholars (including myself) find unacceptable. By offering an interpretation of the *NE* based on Aristotle's views about biological development or "natural production," I show that this view is incorrect and rests on a misconception about the nature of moral virtue.

As I show, morally virtuous activity is indeed a substantive component of flourishing. In fact, morally virtuous activity is a *developmental prerequisite* for contemplative activity. One cannot engage in the latter without possessing the former. Scholars seem to overlook the fact that the ability to engage in excellent contemplative activity is *developed*, and therefore depends upon certain "inner resources," both intellectual and affective, for its realization. Moral virtue implies the presence of these inner resources, as I will show. I will also show that this interpretation of the *NE* reveals that Aristotle intended the *NE* to function as a *transformational work* that helps the pupil to engage in the actions by which contemplative excellence is developed. Aristotle's primary aim in the *NE* is to promote the moral development of his students.

The second aim of this book is to show that far from being outdated, Aristotle's view of flourishing is compelling and defensible even within a liberal state that allows its citizens to pursue their own conceptions of "the good." To that end, I show how the fundamental tenets of my developmentalist interpretation can be used to ground a modern view of flourishing which acknowledges the importance of individual autonomy. Modern virtue theory seeks a return to Aristotle's project by basing a view of flourishing on a scientific understanding of human nature. An emerging movement within psychology known as "positive psychology" offers such an understanding of human nature, and therefore should not be ignored by scholars working within virtue theory. Within positive psychology, one finds a wealth of information about the inner resources and virtues that underlie human flourishing, optimal experience, and well-being.

I focus on a specific domain within positive psychology known as "self-determination theory." Self-determination theory is of particular importance because it not only embraces some of Aristotle's fundamental insights about flourishing, but also recognizes that individual autonomy is a fundamental component of human flourishing. Thus, self-determination theory can be used to ground a modern virtue theory that captures the fundamental tenets of Aristotle's ethical theory, whilst acknowledging the centrality of individual autonomy to flourishing. Aristotle believes that the state's primary purpose is to cultivate the specific virtues that are required for flourishing. As I will show, this view is completely consistent with a regime that allows its citizens to pursue their own conceptions of the good.

The first three chapters of the book are devoted to my developmentalist reading of the *NE*. Chapter 1 provides an exposition of the function argument and some of the famous interpretative problems that surround this argument. In Chapter 2, I show how Aristotle's views on natural production (i.e., biological development) illuminate the moves that he makes in the function argument. Chapter 3 is devoted to a discussion about how ethical virtue or *phronêsis* is a developmental prerequisite of excellent contemplative activity. One becomes an excellent contemplator only if the right motivational and epistemological preconditions are present, and ethical virtue implies both.

Chapters 4 through 6 involve an application of my developmentalist interpretation to a conception of flourishing in which autonomy figures prominently. The basic idea here is that autonomy is an essential component of flourishing, that it is a *developed ability*, and that one develops it in much the same way as one develops the Aristotelian virtues. As mentioned above, this part of the argument makes an appeal to self-determination theory, and so in Chapter 4, I discuss this theory and its affinities to Aristotle's ethics. In Chapter 5, I discuss the concept of *self-concordance*, which underlies the notion of autonomy that one finds within self-determination theory. A developed ability, self-concordance only comes about if certain inner resources or virtues are present. Chapter 5 involves a discussion of these inner resources and virtues. Finally, in Chapter 6, I discuss several programs which cultivate the modern virtues. As we shall see, rather than being incompatible with liberalism, these programs bolster and support its most cherished value of individual autonomy.

A number of individuals helped in the completion of this book and deserve express thanks. Nicholas D. Smith encouraged and inspired me to develop my developmentalist reading of the *NE* when I was his student at Michigan State University. My dear friend, Michael Russo, helped me to clarify and sharpen many of the claims made herein and, as always, provided me with loving support throughout. My students at Central Michigan University, especially Nicholas Brousseau, Amanda Curler, Emily Hagen, and Sandra Joy Russell, were invaluable in helping me to clarify, distill, and cull. Stephen

Didomenico, a former student and now a colleague, was an inspiration and fabulous sounding board throughout. I also thank my editor at Continuum, Sarah Campbell, for providing patience and gentle guidance. Finally, I am deeply grateful to my beloved and magical husband, Jeffrey Wigand, who was selflessly driven to provide an ecology that was full of nurture, love and encouragement, and who fully respected the writing (and rewriting) process. I cannot thank him enough and it is to him that I dedicate this book.

Abbreviations and Notes on Translations

DA	*De Anima*
DMA	*On the Motion of Animals*
GA	*Generation of Animals*
Mem.	*On Memory*
Met.	*Metaphysics*
NE	*Nicomachean Ethics*
PA	*Parts of Animals*
Phys.	*Physics*
Pol.	*Politics*
Post. An.	*Posterior Analytics*
Rhet.	*Rhetoric*
Top.	*Topics*

Unless otherwise indicated, translations are from Jonathan Barnes, ed., *The Complete Works of Aristotle,* the revised Oxford Translation, vols. 1 and 2 (Princeton: Princeton University Press, 1984).

Translations from the *Nicomachean Ethics* are from David Ross' Oxford Translation, reissued as an Oxford World Classics paperback, revised by J. L. Ackrill and J. O. Urmson (NY: Oxford University Press, 1998).

Translations from *Generation of Animals* and *Parts of Animals* are from A. L. Peck's Loeb Classical Library Translations (Cambridge, Mass: Harvard University Press, 1942 and 1937 respectively).

Visit http://www.aristotlesethics.com for further resources regarding this book, as well as easy access to the web links mentioned in the notes.

Chapter 1

The Intellectualism Debate

Introduction

Aristotle's *Nicomachean Ethics* (*NE*) is devoted to the topic of the human good or *eudaimonia*. Although *eudaimonia* is typically translated as "happiness," it is more accurately defined as *flourishing*. Unlike the term "happiness," which denotes an emotional state that comes and goes, flourishing denotes something deeper, more permanent. Think: *awakening*, *enlightenment*, or *self-actualization*. These terms rather than "happiness" more accurately capture the essence of *eudaimonia*. Whatever the preferred label, the important point is that Aristotle's *NE* is concerned not merely with cheer and joy, but with something much more intense, serious, and meaningful. Henceforth, I use the word "flourishing" to refer to *eudaimonia*.

There is much scholarly debate regarding Aristotle's conception of flourishing. One point on which scholars do agree, however, is that Aristotle identifies flourishing with some sort of *activity*—a *doing something*. Pause for a moment and consider this claim. The view that flourishing consists in the performance of an activity departs from some notions in which flourishing is equated with leisure and *in*activity. In the popular 1999 film *Office Space*, for example, a view of flourishing as inactivity is expressed by the main character Peter Gibbons, who, at one point, suggests that flourishing is "doing nothing."[1] Aristotle could not disagree more. Flourishing consists in an activity of some sort. And although Aristotle gives us a big hint that *reason* is involved in this activity, scholars disagree about the *specific type* of rational activity with which Aristotle identifies flourishing. Being mindful of the fact that there are different varieties of "rational activity" is absolutely crucial for understanding Aristotle's notion of flourishing, and the scholarly debate surrounding it.

For instance, some scholars believe that Aristotle identifies flourishing with a single, solitary, and highly cerebral rational activity—contemplative or philosophical activity. This group of scholars endorse what is referred to as an "intellectualist" reading of the *NE* because on this interpretation, Aristotle equates flourishing with the intellectual activity of contemplation.[2] Other scholars reject this view and instead argue for an "inclusivist" reading of the *NE*.[3] According to this view, Aristotle endorses a complex view of flourishing which includes a *variety* of rational activities. Contemplation is one among these activities, but it "shares the stage" with another important rational activity, namely *morally virtuous activity*.

Now, given that I have just described morally virtuous activity as a type of rational activity, you may be a bit surprised. Although regarding moral behavior as a certain type of rational activity is commonplace among academic philosophers, the average person does not view moral behavior as a type of rational activity. To some, moral behavior flows from one's subjective feelings. To others, moral behavior is equivalent to complying with the precepts of one's religion. And many people understand moral behavior in terms of altruism or similar notions of heartfelt "other regard." Aristotle's notion of morally virtuous activity departs from these notions, however. We will discuss this in more detail later, but for now, keep in mind that Aristotle views morally virtuous activity as a type of rational activity that is different from the "highbrow" activity of contemplation. Thus, the debate between intellectualist and inclusivist interpretations of the *NE* boils down to whether Aristotle identifies flourishing with one highly cerebral rational activity, or whether flourishing includes morally virtuous activity.

A crucial piece to this puzzle is the so-called "function argument" wherein Aristotle articulates his view of flourishing. Aristotle provides his definition of flourishing at the *conclusion* of this argument. We are told there that flourishing is *excellent* rational activity. Actually, the conclusion is a tad more complicated than this. It reads: "[flourishing] is excellent rational activity and if there are several then it is the best and most *teleios*." *Teleios* is an adjective. And because it can be interpreted either as "complete," "perfect," or "final," I am leaving the term untranslated for now. When Aristotle says that flourishing is the *most teleios* excellent rational activity, he is using the *superlative* adjectival form. If you have forgotten elementary grammar, the *superlative* of "good" is "best," whereas the *comparative* of "good" is "better." Here are some more examples:

Adjective	**Comparative form**	**Superlative form**
Good	Better	Best
Tall	Taller	Tallest
Smart	Smarter	Smartest

Typically, when someone uses the superlative form, he intends to "single out" an individual from among others of its kind. When I say "Stephen is the smartest student in the class," I single out Stephen from among other students. This is important for understanding the scholarly debate about Aristotle's conception of flourishing. Because the superlative of *teleios* (which, in Greek, is *teleiotatos*) appears in the conclusion of the function argument, this leads some scholars to conclude that Aristotle intends to single out a specific rational activity from among others.

Importantly, Aristotle discusses three different kinds of excellent rational activities in the *NE:* (1) the activity of the virtues of character; (2) the activity of practical wisdom or *phronêsis;* and (3) contemplative activity or the activity of philosophical wisdom (*sophia*). Keep in mind that this list is not just a list of rational activities but of *excellent* rational activities. The activity of *phronêsis* is excellent practical reasoning whereas the activity of *sophia* is excellent contemplation. In Book VI, Aristotle claims that the activity of the virtues of character and the activity of practical wisdom are inter-entailing (*NE* 1144b). In other words, one does not possess the virtues of character unless one has *phronêsis,* and one does not have *phronêsis* unless one possesses the virtues of character. Because these two excellent rational activities are inter-entailing, I use the term "ethically virtuous activity" to refer to their combination. The table below summarizes Aristotle's taxonomy of excellent rational activities:

	Varieties of excellent rational activity	
1	Activity of the virtues of character	} **Ethically virtuous activity** (1–2)
2	Activity of practical wisdom (*phronêsis*)	
3	Contemplative activity (*sophia*)	

Here's the fundamental question that generates scholarly debate: if flourishing is "excellent rational activity," then to *which* of these activities is Aristotle referring in the conclusion of the function argument? Is he singling out just one, as his use of the superlative adjectival form suggests? Or does Aristotle intend to include both ethically virtuous *and* contemplative activity in his definition of flourishing?

My aim in what follows is to show that the correct answers to these questions must acknowledge Aristotle's views about the biological development of human beings. Importantly, biological development is a type of *production* or *making*. The complex biochemical activity behind the development of a fertilized egg into an embryo, from an embryo to a fetus, from a fetus to an infant, and so on, is a natural process of making—the product in this case being a human being. As we know, once the sperm fertilizes the egg, this "making" occurs quite mechanically—pursuant to the instructions of DNA and RNA—whether we like it or not. And although Aristotle does not talk about DNA and RNA *per se*, his views about the mechanisms involved in natural production have deep affinities to contemporary molecular biology.

How is this relevant to Aristotle's view of flourishing? In my view, the function argument, and hence Aristotle's account of flourishing, can only be properly understood if read in connection with Aristotle's views on natural production and biological development. Human beings are natural products and Aristotle's view of human flourishing is profoundly influenced by this fact. Recent work on the topic has indeed recognized the importance of human nature to Aristotle's account of flourishing.[4] However, no recent commentator has fully appreciated the relevance of Aristotle's natural philosophy to his moral philosophy, and consequently, to his view of *eudaimonia*. Aristotle's claim that the human soul is the form of the body (*DA* 412a), his view that the form of a natural body is subject to the natural processes that govern its development or actualization (*PA* 639b–640a), and his view that what is peculiar (*idion*) to a creature is formed *last* in that creature's development (*GA* 736b), are fundamental to Aristotle's view of flourishing, as I will show.

In order to substantiate these claims, I will first explicate Aristotle's function argument. Aristotle uses this argument not only to advance his own theory of flourishing, but also to demonstrate the inadequacy of rival theories of flourishing. After explicating this argument, I will discuss the two rival interpretations of Aristotle's view of flourishing: intellectualism and inclusivism. Doing so will motivate a developmentalist interpretation of Aristotle's view of flourishing that I take up in the following chapter. Let's begin, then, with an explication of Aristotle's function argument.

The Function Argument

The function argument (*NE* 1097b–1098a) is Aristotle's argument for his view of human flourishing. This argument is complex and consists of a number of moves including several analogical arguments. I will explain the argument by focusing on one specific move or "conceptual chunk" at a time. Then, we will see how all of these moves work together to ground Aristotle's unique view of flourishing.

According to the function argument, the human good can be defined through the *ergon* of man (*NE* 1097b). The word "*ergon*" is usually translated as "function," or "characteristic work." Sometimes this term is translated as "purpose."[5] For the purpose of explicating the function argument, I leave the term "*ergon*" untranslated.

Aristotle provides an analogical argument for his claim that the human good is determined by man's *ergon*:

> [A clearer account of flourishing] might perhaps be given if we could first determine the *ergon* of man. Just as for a flute player, a sculptor, or any artist, and in general, for all things that have an *ergon* or activity, the good and the 'well' is thought to reside in the *ergon*, so it would seem to be for man, if he has an *ergon*. (*NE* 1097b)[6]

Aristotle makes two important claims here. First, he claims that the good or flourishing of an artist/craftsman resides in the purpose or function of his respective art. A good or flourishing flautist is one who performs her function (e.g., playing the flute) well; a good or flourishing sculptor is one who sculpts well; and a good or flourishing football player is one who plays the game well. The second, more contentious claim is that man—considered apart from any specific trade or pursuit like flute playing, sculpting, or football—has a function. In fact, Aristotle appears to conclude this fact about man from some strange evidence. Aristotle considers the various functions of the trades and crafts in which man can engage, and concludes from this fact that man *himself* has a function. So the argument looks like this:

1. The *ergon* of a flute player is playing the flute.
2. A flourishing/good flute player is one who performs her *ergon* well.
3. The *ergon* of a sculptor is sculpting.
4. A flourishing/good sculptor is one who performs his *ergon* well.
5. Therefore, a flourishing/good artist is one who performs the *ergon* of one's art well.
6. Therefore, if a human being has an *ergon*, a flourishing/good human being is one who performs that *ergon* well.

To construct his theory of flourishing, Aristotle begins by zooming in on the crafts and observing something about the good or "flourishing" of the individuals practicing those crafts. He then analogizes craftsmen to human beings, arguing that if the former have an *ergon* and a function, then so do the latter.

Goods and goals

The conclusion of the above argument connects the *ergon* of a human being with the good of a human being. Importantly, the word "good" for Aristotle denotes something rather different from the conventional understanding of this term:

> Let us return to the good that we are seeking and ask what it can be. It seems different in different actions and arts; it is different in

> medicine, in strategy, and in the other arts likewise. What then is the good of each? Surely that for whose sake everything else is done. In medicine this is health, in strategy victory, in architecture a house, in any other sphere something else, and in every action and pursuit the end. (*NE* 1097a)

For Aristotle, the word "good" does not denote something like "moral value." The good is "the that for whose sake everything is done." Aristotle, therefore, uses the word "good" to denote an ultimate end or *goal*. Thus the good of an art is the *goal* or end of that art and "the that for whose sake everything else is done." The good of medicine is its *goal*: health. The good of strategy is its *goal*: victory. The good of architecture is its *goal*: a house. Therefore a "good doctor" is simply one who achieves the goal of health, and achieves it well. Aristotle's notion of "the good" is much *narrower* than our notion of this term. Aristotle's notion is concerned with goals, no more, no less.

The human good (i.e., goal)

From his observations about the good (i.e., goals) of artists and craftsmen, Aristotle draws conclusions about the good and flourishing of human beings. On the face of it, such an approach seems problematic. Aristotle draws an analogy between craftsmen (musicians, sculptors) and human beings, and it is not at all clear why craftsmen have any significant similarity to human beings. After all, assuming that they have some sort of professional liberty, artisans and craftsmen seem to *choose* their trade, their end and their "purpose." And it is not at all clear that a human being, *qua* human being, has some sort of natural or inborn end. The existentialist philosopher Jean Paul Sartre, for example, claims that "man exists, turns up, appears on the scene, and, only afterwards, defines himself" and "[m]an is nothing else but what he makes of himself."[7] In Sartre's view, therefore, since there is no specific end that constrains the human being, a human being is most *un*like a craftsman. The actions of the craftsman are constrained by the end of his craft, but the actions of human beings are constrained by no such end, in Sartre's view. Thus, Aristotle's analogy between craftsmen and human nature seems weak.

Nevertheless, Aristotle continues his argument as if he has said nothing special. To support his claim that man indeed has an *ergon*, Aristotle not only uses evidence from the crafts (yet again), but he also draws evidence from human anatomy. Here is the next part of the argument:

> Have the carpenter, then, and the tanner certain functions or activities, and has man none? Is he born without an *ergon*? Or as eye, hand, foot and in general each of the parts has an *ergon*, may one lay it down that man similarly has an *ergon* apart from all these? (*NE* 1097b)

Once again, Aristotle analogizes craftsmen (the carpenter, the tanner) to human beings. And to lend support to the claim that man does indeed have an *ergon*, Aristotle next asks us to consider body parts. Here is the subsequent part of the argument:

7. The carpenter and the tanner each have their own *ergon*.
8. The eye, hand, foot and other human body parts each have their own *ergon*.
9. Therefore, man has his own *ergon*.

Essentially, Aristotle's position is this: "craftsmen and body parts each have their own *ergon*, so why wouldn't man?" But again, the similarities between the crafts, body parts, and human beings seem weak. Why, after all, does Aristotle think that man has an *ergon*, just because craftsmen do? And is Aristotle claiming that a human being has an *ergon* just because his component parts do? Terence Irwin puts the problem well:

> When Aristotle offers to say what happiness is, he refers to the "function" or "characteristic activity" (*ergon*) peculiar to human beings: just as a hammer or a leg has a function, a living organism has one . . . This argument is not easy to accept. Is the inference

> from artifacts and organs to human beings secure, or does it rely on the disputable assumption that a human being is an artifact, or else the limb of a larger organism?[8]

In addition to the fact that Aristotle's comparison between human beings and craftsmen/artists is problematic, the argument also runs awry because it appears to commit the fallacy of composition. For in premises 8–9 above, Aristotle seems to accept the general premise that "if the parts of X each have an *ergon* or function, then X itself has an *ergon* or function." But it simply does not follow that if P is predicated of each of the parts of X, then P can also be predicated of X itself. The atoms comprising Socrates cannot be seen with the naked eye, but this does not imply that Socrates cannot be seen with the naked eye. So, the fact that the parts of the human body have their respective functions, does not imply that the human body, as a whole, has a function.

Despite the difficulties surrounding premises 7–9, Aristotle proceeds as if he has said nothing remarkable. And after establishing that man has an *ergon*, Aristotle next claims that the *ergon* of a human being can be discovered if one discovers what is *idion* to human beings (*NE* 1097b):

> What then can [the *ergon* of man] be? Life seems to belong even to plants, but we are seeking what is peculiar (*idion*) to man. Let us exclude, therefore, the life of nutrition and growth. Next there would be a life of perception, but *it* also seems to be shared by the horse, the ox and every animal. There remains, then, an active life of the element that has a rational principle. (*NE* 1097b–1098a)

According to Aristotle, the *ergon* or function of man can be determined if one discovers what is *idion* to human beings. Most scholars interpret "*idion*" as "peculiar."[9] And here again another problem appears to present itself. If Aristotle is claiming that the *ergon* of a human being is determined by what is peculiar to a human being, then the connection between a thing's *ergon* and its flourishing is tenuous. First, why is it that man's *ergon* is discovered by determining what is peculiar to human being? Second, why should what is "peculiar" *matter* to human flourishing? And finally, why does Aristotle select "rational activity" as that which is peculiar to man? Indeed, there are many characteristics that are peculiar to human beings *other than* rational activity such as prostitution and burning ants with a magnifying glass, to name but a few. So why doesn't Aristotle single out *these* activities in his account of flourishing? Aristotle appears to provide no answer to these fundamental questions. His argument here is summed up as follows:

10. The *ergon* of man can be discovered by determining what is *idion* to man.
11. Activity in accordance with the rational element is *idion* to man.
12. Therefore, the *ergon* of man is activity in accordance with the rational element.

After stating that the *ergon* of man is activity in accordance with the rational element, Aristotle then makes the important claim that the rational element in man is twofold:

13. One part of the element that has a rational principle has such a principle in the sense of being obedient to one, the other in the sense of possessing one and exercising thought.

After establishing that there are two parts to the rational element of the soul, Aristotle then reminds us that since flourishing consists in performing the *ergon* well, and since the *ergon* of man is rational activity, then human flourishing consists in the best and *teleiotatos* rational activity. So Aristotle concludes the function argument with the following claim:

14. [human flourishing is] activity of the [rational] soul exhibiting excellence, and if there is more than one excellent [rational activity], then it is the best (*aristos*) and *teleiotatos*.

I am leaving "*teleiotatos*" untranslated because, as mentioned above, the word could be translated either as "most perfect," as "most complete," or as "most final." And as we shall see, scholars disagree about how "*teleiotatos*" should be rendered. Aristotle's function argument can be summarized as follows:

1. The *ergon* of a flute player is playing the flute.
2. A flourishing/good flute player is one who performs her *ergon* well.
3. The *ergon* of a sculptor is sculpting.
4. A flourishing/good sculptor is one who performs his *ergon* well.
5. Therefore, a flourishing artist/craftsman is one who performs the *ergon* of one's art/craft well.
6. Therefore, if a human being has an *ergon*, a flourishing/good human being is one who performs that *ergon* well.
7. The carpenter and the tanner each have their own *ergon*.
8. The eye, hand, foot, and other human body parts each have their own *ergon*.
9. Therefore, man has his own *ergon*.
10. The *ergon* of man can be discovered by determining what is *idion* to man.
11. Activity in accordance with the rational element is *idion* to man.
12. Therefore, the *ergon* of man is activity in accordance with the rational element.
13. There are two parts to the rational element.
14. Therefore, human flourishing is excellent activity of the rational element, and if there is more than one rational element, then human flourishing is excellent activity of the rational element that is best and *teleiotatos*.

The qualification that Aristotle makes at the conclusion of the function argument, namely, that the human good is excellent activity in accordance with the best and *teleiotatos* rational element, refers back to his remark that the rational element has two parts (NE 1098a). Thus the claim seems to be that flourishing is excellent activity in accordance with the best and *teleiotatos* of these two parts.

The Difficulty of Book X

So far we have laid out Aristotle's function argument and have mentioned some of the difficulties involved therein. However, in addition to the seemingly problematic moves that Aristotle makes in the function argument, his remarks about contemplative reason in Book X, the final book of the *NE*, further complicate the matter. Recall our earlier discussion above about the three different types of excellent rational activity discussed in the *NE:* (1) the activity of the virtues of character; (2) the activity of practical wisdom; and (3) the activity of contemplative or philosophical reason.

1	Activity of the virtues of character	} **Ethically virtuous activity**
2	Activity of practical wisdom (*phronêsis*)	
3	Contemplative activity of philosophical wisdom (*sophia*)	

The earlier books of the *NE*, especially books I through VIII, contain detailed discussions about ethically virtuous activity. Aristotle indeed spends the bulk of the *NE* discussing ethically virtuous activity, whereas the discussion of contemplative activity occupies only a small portion of the *NE*. Given that Aristotle spends the bulk of the *NE* discussing ethically virtuous activity, one is naturally inclined to believe that ethically virtuous activity would be central to Aristotle's account of flourishing. However, Aristotle's remarks in Book X, the final book of the *NE*, seem to suggest otherwise. For in Book X, Aristotle claims that the best activity is not ethically virtuous activity, as one might expect, but is contemplative activity—activity that proceeds from the highest virtue, *sophia*:

> If happiness is activity in accordance with virtue, it is reasonable that it should be in accordance with the highest virtue; and this will be that of the best thing in us . . . the activity of this in accordance with its proper virtue will be [*teleios*] happiness. That this activity is contemplative we have already said. (*NE* 1177a)

Later, at X.8, Aristotle explicitly states that the life that includes only ethically virtuous activity is inferior to a life that contains contemplative activity:

> But in a secondary degree the life in accordance with [ethically virtuous activity] is happy . . . the philosopher, more than any other, will be happy. (*NE* 1178a–1179a)

Book X seems to be out of place in the *NE*. The earlier books suggest that ethically virtuous activity is an integral component of flourishing, whereas the explicit statements about the supremacy of contemplative activity seem to negate this view. Indeed, given that Aristotle uses superlative adjectives, he seems to be singling out a specific rational activity, one that is "best" and "*teleiotatos*." Does he believe that ethically virtuous activity is the best and *teleiotatos*? Or does he believe that the best and *teleiotatos* activity is contemplative? Or, perhaps, to consider another possibility, Aristotle believes that the best and *teleiotatos* activity is a combination of both ethically virtuous and contemplative activities. Aristotle's function argument seems to involve a thicket of difficulties. Not only do his moves within the function argument seem unacceptable, but there is also the tension between Book X and the earlier books of the *NE*.

Scholars have been particularly frustrated by the fact that Aristotle is unclear about which rational activity he believes is best and *teleiotatos*, and there is notorious debate about how the conclusion of the function argument should be interpreted. Should the conclusion of the function argument be interpreted in light of Book X, or should Book X be eschewed when interpreting the function argument? There are indeed scholars who endorse the latter alternative, and argue that certain portions of Book X were not originally part of the *NE*.[10] Such an account at least acquits Aristotle of outright inconsistency, as it places the blame on some careless editor of Aristotle's corpus. Nonetheless, even if portions of Book X were not originally part of the *NE*, the function argument in Book I remains problematic due to its dependence on seemingly weak analogies and specious logical moves.

Other scholars advance less charitable accounts of the relationship between Book X and the rest of the *NE*. According to this school, Aristotle actually contradicts himself in the *NE*. W. F. R. Hardie and John Cooper, for instance, argue that Aristotle identifies the best and *teleiotatos* activity with ethically virtuous activity in the first nine books of the *NE*, and identifies the best and *teleiotatos* activity with contemplative activity in the final book of the *NE*. Such an account interprets the *NE* as containing a flagrant contradiction.[11]

There are, however, many scholars who are more charitable with Aristotle, arguing that he presents a unified and coherent account of the best and *teleiotatos* activity in the *NE.*[12] Yet, there is disagreement even among the scholars who provide this more charitable interpretation of Aristotle. Essentially, the disagreement concerns whether Aristotle identifies the best and *teleiotatos* activity with the sole activity of contemplation, or whether he believes that the best and *teleiotatos* activity consists in a combination of ethically virtuous and contemplative activities. *Intellectualists* endorse the former alternative, whereas *inclusivists* endorse the latter. I think that both of these interpretations are problematic because they fail to appreciate that Aristotle's views about biological development are fundamental to the logic of the function argument. After briefly describing intellectualism and inclusivism, I will outline an interpretation of the function argument that is grounded in Aristotle's views about the biological development of human beings. I call this approach to Aristotle's ethics *developmentalism.* A developmentalist interpretation of the *NE* illuminates both the logic of the function argument and the relationship between ethically virtuous and contemplative activity, as we shall see.

Intellectualism

According to some scholars, Aristotle identifies the best and *teleiotatos* activity with contemplative activity, and hence believes that there is but one fundamental ingredient to flourishing, namely, contemplative activity.[13] These scholars endorse an intellectualist interpretation of Aristotelian flourishing. Intellectualists claim that all other activities, including ethically virtuous activity, are valuable only because they help to promote contemplative activity. In order to support their position that the best and *teleiotatos* rational activity is contemplative activity, intellectualists interpret *teleiotatos* as "most perfect," and argue that Aristotle's use of superlative adjectives in the conclusion of the function argument signifies his intent to identify flourishing with a single, solitary activity. And given Aristotle's remarks about the supremacy of contemplative activity in Book X, it makes perfect sense to identify flourishing as contemplative activity, no more, no less.

Although intellectualism accords well with Aristotle's remarks in Book X, many scholars reject this reading because of its rather unpalatable consequences. After all, intellectualism implies that it is possible for a flourishing life to be devoid of ethically virtuous activity. Thus, intellectualists claim that since ethically virtuous activity is not a fundamental component of a flourishing life, then it is possible for a flourishing individual to shirk ethically virtuous activity.[14] Richard Kraut, for instance, claims that in some cases, contemplative activity can and should be maximized by abandoning ethically virtuous activity.[15]

> [I]f you manage to escape detection, then certain injustices might bring you more contemplation than less. For example, if you ignore the needs of your parents, and no one finds out about this neglect, then your standing in the community will not suffer, and you will have more time for theoretical activity. My interpretation does not attribute to Aristotle the assumption that such cases cannot occur.[16]

Many have found intellectualism unacceptable since it implies that one can flourish while ignoring the needs of one's family, friends, and community. Scholars simply cannot accept the idea that Aristotle believes that human flourishing is solely concerned with contemplative activity, and only acknowledges the value of social, political, and family life insofar as these activities promote contemplative activity. If this were true, then Aristotle would be arguing for a life that, in our own day, would be looked upon with disdain. Thus, many scholars reject an intellectualist interpretation of the *NE* since it saddles Aristotle with a rather impoverished view of human flourishing.

Inclusivism

Because many scholars find an intellectualist interpretation of the *NE* unacceptable, they embrace an alternative *inclusivist* interpretation.[17] Scholars who endorse inclusivism reject the view that flourishing consists in contemplative activity alone, and argue instead that flourishing is a complex, multidimensional end composed of several rational activities, namely, ethically virtuous and contemplative activities. Thus, inclusivists identify the best and *teleiotatos* activity with a *combination* of activities. On such a view, the flourishing individual is *necessarily* ethically virtuous, since such activity is an essential component of flourishing.[18] In contrast to intellectualists, inclusivists interpret *teleiotatos* as "most complete."[19] Thus, the best and *teleiotatos* rational activity is the most complete, all inclusive activity.

Although all inclusivists agree that both ethically virtuous and contemplative activity are components of human flourishing, some believe that theoretical and ethically virtuous activity are *equally* desirable, whereas others believe that theoretical activity is *more desirable* than ethically virtuous activity.[20] I call the former group *simple inclusivists*, and the latter group *moderate inclusivists*.[21]

There are, then, at least three different positions within the intellectualism debate. Interpreting "*teleiotatos*" as "most perfect," *intellectualists* claim that flourishing consists in contemplative activity alone, that ethically virtuous activity is not a component of flourishing, and that it is valuable only to the extent that it promotes contemplative activity. *Simple inclusivists* claim that contemplative and ethically virtuous activity are both equally important components of flourishing. Finally, *moderate inclusivists* agree that flourishing includes ethically virtuous and contemplative activity, but add that contemplative activity is more desirable than ethically virtuous activity. Both groups of inclusivists interpret "*teleiotatos*" as "most complete." The table below summarizes these different interpretive positions:

	Flourishing consists in contemplative activity alone?	**Ethically virtuous activity an essential component of flourishing?**	**Ethically virtuous activity and contemplative activity equally desirable?**	**Translation of "*teleiotatos*"?**
Intellectualist	YES	NO	NO	Most perfect
Simple inclusivist	NO	YES	YES	Most complete
Moderate inclusivist	NO	YES	NO. Contemplative activity is more desirable.	Most complete

Developmentalism

In the following chapter, I argue for an interpretation of the function argument that is grounded in Aristotle's views about the natural or biological development of human beings. The intellectualist and inclusivist interpretations of the function argument are inadequate because both views fail to recognize that Aristotle's view of flourishing is informed by his views about biological development. In *Generation of Animals, Parts of Animals* and *Physics,* Aristotle repeatedly discusses how nature brings about some product or process, for the sake of some subsequent stage of development (*GA* 743a, *PA* 639b–640a, *Phys.* 199a). Additionally, in these works of natural philosophy Aristotle explicitly states that the *idion* or characteristic function of a species plays an important role in the biological development of that species. When we understand that Aristotle's views about natural development are fundamental to the logic of the function argument, a reading of this argument presents itself which illuminates the relationship between ethically virtuous and contemplative activity. It is to this developmentalist reading of the function argument that we now turn.

Chapter 2

A Developmentalist Interpretation of the Function Argument

Introduction

As we discussed in the Chapter 1, there seem to be a number of problems with the very logic of the function argument. In addition to the problems presented by Book X, the function argument seems to be infected by weak analogies and other fallacious moves. However, once we understand the biological basis of the function argument, the alleged difficulties with Aristotle's function argument vanish.

I will begin by discussing Aristotle's theory of causality and will then turn to the corresponding notion of *hypothetical necessity*. The doctrines of final causality and hypothetical necessity are fundamental to Aristotle's theory of *production* or making regarding both man-made artifacts such as tools and houses, and natural organisms such as dogs and human beings. However, I believe that Aristotle's notions of final causality and hypothetical necessity are also essential to understanding his views about human flourishing. In fact, I believe that these notions are so fundamental to Aristotle's theory of flourishing that I interpret "*teleiotatos*" as "most final." In my view, Aristotle alludes to the notion of final causality at the conclusion of the function argument.

Importantly, however, there are two dimensions to Aristotle's view of final causality, both of which are important to understanding his view of flourishing. One dimension concerns final causality in the natural, observable world. I call this the *objective* dimension of final causality. The other dimension of final causality—equally important to Aristotle's view of flourishing—concerns the internal, psychological components in an individual's goal directed actions, and, in particular, the individual's *desires*. I call this the *desiderative* dimension of final causality. A proper understanding of Aristotle's view of flourishing requires an acknowledgement of both of these dimensions of final causality. Aristotle's view of flourishing concerns the relationship between these two dimensions, as we will see.

I begin by first discussing how final causality and hypothetical necessity operate in man-made production. Keep in mind that final causality and hypothetical necessity operate in *both* man-made production and in natural production. And although it is *natural* production that is essential to understanding Aristotle's view of flourishing, I begin by focusing on man-made production for two reasons. First, because man-made production involves the goal directed actions of the craftsman, it allows us to understand the motivational system of the craftsman and therefore affords an opportunity to understand the desiderative dimension of final causality. Second, the focus on man-made production provides a simpler, more intuitive way of understanding the operation of final causality and hypothetical necessity. Understanding how final causality and hypothetical necessity operate in the domain of man-made production makes it easier to understand how they work in the domain of natural production and biological development.

Causality in Man-made Production

Aristotle's theory of causality is fundamentally a theory about production and making. Ultimately, Aristotle is concerned with the elements necessary for producing some sophisticated entity like a hammer from raw materials such as wood and steel. In this, he departs from narrow conceptions of causality such as the one discussed by David Hume. In discussing causality, for instance, Hume focuses on the behavior of billiard balls, noting that we expect the motion of one to cause the motion of another upon impact.[1] Famously, Hume used such examples to show that such causal expectations are not grounded in absolute certitude. This issue did not concern Aristotle. Again, Aristotle is concerned with *making*. And of course, the causal effect that one billiard ball has upon another is an altogether different matter from the question of how both billiard balls came into being. Contrast the question "what happens when the cue ball hits the eight ball?" with the question "how does one go about making a cue ball and an eight ball?" Aristotle is concerned with the latter question.

His theory of causality addresses the processes involved in the making of the billiard balls, rather than on the effects that billiard balls can have on one another after they are made.

Aristotle's theory of causality is discussed in Book II of the *Physics.* There, he describes four different causes: the material, efficient, formal, and final (*Phys.* 194b–195a). The *material* cause is the matter from which a product is made; the *efficient* cause is that which sets man-made production in motion; the *formal* cause is the essence of the product, what the product is. The *final* cause is the "that for the sake of which" something is made or done.

Because Aristotle's theory of causality is a theory about the different factors or "causes" that go into the production or the making of something, consider, for example, the making of a pen. In order to make a pen, one first has to find the right kind of material—at the very least one needs a firm substance that does not dissolve in one's hand when it is held. Thus, pens are not made out of water, dirt, or flour, but are made out of materials like plastic or some sort of metal like silver. Aristotle refers to the material out of which a thing is made as its *material cause.*

Of course, one does not have a pen once the right kind of material is found. One then needs to manipulate that material to produce the pen—it has to be shaped, drilled, sanded, and so on—the material has to be *set in motion.* Aristotle refers to the agent that sets the material in motion as the *efficient cause.* Let's suppose that Namiki is a pen maker who has a custom order for a sterling silver pen. Namiki is therefore the efficient cause of this pen. He sets the process of pen-making in motion. From the buying of the silver, to the melting of it in a crucible, to the pouring of into a mold, Namiki is the force that drives the process.

Note that Namiki does not buy the silver on a whim. Nor does he manipulate the silver for the fun of it. Rather, he does these things *for the sake of* making a pen. Every step, from the acquiring of the material, to the melting, the molding, and the buffing is done for the sake of producing a pen. Namiki does not shape and buff simply for fun. There is a "final cause"—a goal—directing Namiki to move his hands this way and that. Aristotle refers to the end or goal of a process as its *final cause.* As mentioned earlier, Aristotle's notion of final causality is particularly relevant to the conclusion of the function argument and therefore to his view of flourishing.

The fourth cause of which Aristotle speaks is the "formal cause." The form of a thing is the specific function that the thing is designed to do. Thus, the formal cause of a pen is what the pen is designed to do, namely, to write. Because the formal cause is not realized until the pen is made, Aristotle says that the final cause and the formal cause are identical (*Met.* 1044b).

Aristotle's doctrine of the four causes can be summed up as follows: the "material cause" is the material or "stuff" out of which an object is made; the "efficient cause" is what sets the material in motion; the "final cause" is the goal or end of a process; and the "formal cause" is related to the product's function and what it is designed *to do.* The following table summarizes Aristotle's theory of causality.

Causality in man-made production

Type of cause	**Definition**	**Example**
Material	*The stuff out of which the object is made*	Silver
Efficient	*What sets the production in motion*	Namiki, pen maker
Formal	*What the product is designed to do, its function*	An instrument for writing
Final	*The finished product that performs a specific function*	A silver pen

Hypothetical necessity in man-made production

In addition to his theory of causality, Aristotle's notion of *hypothetical necessity* is essential to understanding his views about production. One could say that hypothetical necessity is the mechanism by which final causality operates.

An hypothetical necessity is simply a condition that must obtain *if* a certain end is to be realized. Thus, *if* one is going to make a silver pen, then silver is necessary for its completion, and *if* one needs silver, then one must go to the market to purchase this material. Thus, purchasing silver is hypothetically necessary for the production of a silver pen, and going to the market is hypothetically necessary for the purchasing of silver. Aristotle describes how hypothetical necessity is involved in the medical, rhetorical, and political crafts:

> [A] doctor does not deliberate about whether he should heal, nor an orator whether he shall convince, nor a statesman whether

> he shall produce law and order, nor does any one else deliberate about his end. Having set the end, they consider how and by what means it is to be attained; and if it seems to be produced by several means they consider by which it is most easily and best produced, while if it is achieved by one only, they consider how it will be achieved by this, and by what means this will be achieved, till they come to the first cause which in the order of discovery is last. (*NE* 1112b)

> The healthy subject, then, is produced as the result of the following train of thought; since *this* is health, if the subject is to be healthy, *this* must first be present, e.g., a uniform state of the body, and if this is to be present, there must be heat; and the physician goes on thinking thus until he reduces the matter to a final step which he himself can take. Then the process from this point onward, i.e., the process towards health, is called a 'making.' (*Met.* 1032b)

Aristotle's notion of hypothetical necessity reveals something crucial about final causes. A final cause is the end or goal of an action. But the doctrine of hypothetical necessity reveals that *some goals are more final than others.* Indeed, some goals are realized solely because they are necessary for the attainment of more final goals. Put bluntly, some goals are also *means.* If Namiki needs to walk to the market to purchase silver (PS), then PS is the final cause of the walk. But although PS is the final cause and the goal of the walk, it is not the *most* final cause of Namiki's actions. PS is done because it is *necessary* for realizing Namiki's ultimate goal of making a pen. PS is a goal *and* a means. In other words, PS is a goal that is a means by which to obtain some further goal (a silver pen). Since some goals are means for further goals, then there is a serial or "nested" order of goals. Some final causes are more "final" than others.

The notion that there are degrees of finality and that there is a serial order of goals and ends, is, I believe, crucial for understanding Aristotle's theory of flourishing. And to see why, let's begin by shifting gears for a moment and discussing the motivational aspects of goal-directed behavior—the inner resources of the craftsman that enable him to undertake orderly sequence of actions necessary for production. As we shall see, there are a number of inner resources that are implicated in the craftsman's goal directed action. I refer to the collection of the inner resources necessary for goal attainment as the *motivational system.* The motivational system involves both *intellectual* components on the one hand, and *desiderative* components, on the other. Let's begin by discussing the intellectual components of the craftsman's motivational system.

The Craftsman's Motivational System

The intellectual components

The above passages that discuss hypothetical necessity indicate that prior to production, a craftsman first engages in practical deliberation about how to produce the product. Before he makes, the craftsman thinks. As Aristotle says, "[t]he healthy subject is produced as the result of the following train of thought: since *this* is health, if the subject is to be healthy, *this* must first be present" (*Met.* 1032b). "Practical deliberation" is not just thinking, it is a *specific type* of thinking—"strategic planning." There is a specific goal—an objective—and strategic planning consists in the discovery and calculation of the best means by which the goal is achieved.

Strategic planning

Aristotle's remarks portray strategic planning as a type of "backwards thinking." The craftsman begins with the goal or end ("since this is health"). From that goal, the craftsman then reasons backwards through a chain of hypotheticals until he reaches a condition that he can produce at once (e.g., rubbing hands to produce heat). In the above passage from the *NE,* Aristotle describes strategic planning as follows: "[craftsmen] consider how [their end] will be achieved by *this,* and by what means *this* will be achieved." Given our discussion above about serial goals and degrees of finality, we can say that each "this" is *both* a necessary condition for the realization of the goal *and* a final cause of some earlier step. Each "this" is both an end and a means by which some further end is achieved.

Thus, Namiki starting from his end and goal, reasons through a chain of hypothetical necessities that is, in effect, the algorithm for the production of the pen ("I want to make a silver pen. So then I need silver. I do not have silver, so I need to go to the market. I have the money. I will then go to the market now.") The sort of thinking that Namiki engages in prior to his journey to the market is no ordinary type of thinking. Namiki is neither daydreaming nor reminiscing about the good old days. He is engaged in practical deliberation and formulating a strategic plan—a blueprint or recipe for making a silver pen. Importantly, the strategic plan reveals the specific order of "mini-goals" that Namiki must realize in order to achieve his ultimate goal of making a pen.

The strategic planning involved in man-made production shows that the craftsman is not just a doer. He is also a thinker and a planner. He engages in strategic planning because it is necessary to realize his end of production. Planning is a type of thinking, a type of rational activity, and therefore is an intellectual component of the craftsman's motivational system.

According to Aristotle, however, the practical deliberation and the subsequent strategic plan for making an artifact proceeds from its form (*eidos*). Form corresponds to the essence and function of a thing (*Met.* 1032b). In the context of production, it is useful to think of the "form" of an artifact as the *knowledge* required to make it. In addition to practical deliberation, form, therefore, is another intellectual component in the craftsman's motivational system that is essential to production.

Form or knowledge

Contrast yourself, who probably does not know how to build an airplane, with someone who does. Because you do not understand how the different parts of the plane contribute to its overall function, and because you further do not understand what is needed for the parts to exist, you would not even know where to start in the deliberative process required to generate the blueprint for its production.

Nevertheless, because the aeronautical engineer understands the complex components and their interrelations needed to make an airplane, he is able to distill or "unpack" its essence or form into its component parts through deliberation, thereby resulting in the chain of hypothetical necessities or "recipe" for its production.

To modern ears, it may sound awkward to speak of the role that "form" plays in production. But the idea is simple. Consider the lyrics from Peter Gabriel's song *Mercy Street*: "*all of the buildings, all of those cars were once just a dream in somebody's head.*" The production of artifacts—buildings, cars—starts with an idea—"a dream in somebody's head." This is essentially what Aristotle means when he claims that form initiates production:

> [N]othing passes from the carpenter into the pieces of timber, which are *his* material, and there is no part of the art of carpentry present in the object which is being fashioned: it is the shape and form (*eidos*) which pass from the carpenter, and they come into being by means of the movement in the material. It is in his soul, wherein is the "form," and his knowledge, which cause his hands (or some other part of his body) to move in a particular way (different ways for different products, and always the same way for one product); his hands move his tools and his tools move the material. (*GA* 730b)

> The active principle then and the starting-point for the process of becoming healthy is, if it happens by art, the form in the soul. (*Met.* 1032b)

Granted, Aristotle's notion of production is a tad more precise than saying that buildings and cars come from a "dream in somebody's head." There is the "dream," and then there is the *deliberation* which produces the recipe for building the "dream." Both practical deliberation *and* the form from which it proceeds are essential intellectual components in the motivational system of the craftsman. And without the craftsman's motivational system, no making or production is possible.

We have been talking about the intellectual components of the craftsman's motivational system. But as mentioned above, the intellectual components are necessary, but not sufficient for production.

In order for Namiki to implement the blueprint for making the pen, he must *act.* He must move his limbs and hands in specific ways that the recipe directs. As Aristotle states in the *De Anima,* something *else* is needed to cause the craftsman to act. After Namiki realizes that he must go to the market to get silver, he must get up, move his body, and make the trek. And Aristotle explicitly states that it is not the knowledge of how to make a pen that causes Namiki to make the trek, but some "other thing":

> And quite generally we observe that it is not the man just possessed of medical skill that cures, there being some other thing, not the knowledge, that is responsible for the production of the cure in accordance with knowledge. (*DA* 433a)[2]

In addition to intellectual and cognitive components of the craftsman's motivational system, *desiderative* components are also crucial. As Aristotle states, "the object of desire is the point of departure for action" (*DA* 433a) and "action cannot occur without desire" (*DMA* 701a). Thus, Namiki's faculty of desire is the some "other thing" needed to implement the blueprint for making a pen.

The desiderative components

A word of caution as we discuss Aristotle's theory of desire. The commonplace view of desire is associated with *wanting* or *yearning* for something pleasurable. For Aristotle, however, "desire" extends beyond the notions of wanting or yearning. And as we shall see, there are some forms of desire which are *not* associated with pleasure. Aristotle thinks of desire as a *causal power.* But recall that Aristotle's view of causality is complex: Aristotle recognizes material, efficient, formal, and final causes. Thus it isn't enough to say that desire has causal power. Given Aristotle's complex view of causality, one also has to ask about the *type* of cause that a desire is. Desire functions as both an *efficient* and a *final* cause. But for simplicity's sake let's use the similar notions of *means* and *end,* notions which Aristotle himself uses in articulating his theory of desire. In fact, Aristotle explicitly distinguishes *desire for the means* from the *desire for an end or goal.*

The former he calls *choice*, the latter he calls *wish* (*NE* 1111b). Indeed, in order for Namiki to actually produce a pen, he must not only desire to bring about the goal of a completed silver pen, but he also must desire to bring about the means of production. "Wish" refers to the former type of desire, whereas "choice" refers to the latter. Let us begin by discussing choice.

Choice

Choice (*prohairesis*) is a specific type of desire; desire not for the end but for the means by which some end is realized (*NE* 1111b). Aristotle claims that choice is *deliberate desire*, that is, a desire that is caused by, and hence emerges from, practical deliberation:

> The object of choice being one of the things in our own power . . . choice will be deliberate desire of things in our own power; for when we have reached a judgement as result of deliberation, we desire in accordance with our deliberation. (*NE* 1113a)

Unlike *simple* or *regular* desire that is concerned with pleasure, choice is caused by practical deliberation and is *not* concerned with pleasure (*NE* 1111b). Choice is an *effect* of practical thinking, albeit an effect with causal power to move the body. Put in the context of Aristotle's theory of causality, choice is an effect that is the efficient cause of the craftsman's goal-directed bodily movement. As Aristotle says, choice is the starting point and the efficient cause of action (*NE* 1139a).

Consider, again, Namiki. Suppose he has been working on his pen all day and then begins to feel hunger pangs. His desire for food is not, of course, the result of deliberation. It just happens to him. If Namiki takes a lunch break and satisfies his hunger, he will feel the pleasure associated with satisfying this desire. Contrast Namiki's desire for food with his desire to go to the market to buy silver. The latter desire does not just "happen" to Namiki, nor does the completion of this task satisfy one of Namiki's urges. The desire to go to the market emerges as part of Namiki's recipe for making the silver pen and hence is part of the means by which this goal is realized. Namiki, of course, *must* perform the necessary means in order to make a pen. Without performing these means, no pen will be produced. Therefore, pen-making implies some inner causal power that drives Namiki to perform these means. The causal power which compels Namiki to perform these means is *choice.* Again, choice is the efficient cause of the craftsman's goal-directed bodily movement.

But now let's talk about that which choice is the effect *of.* Where does choice come from? Practical deliberation. But from what does practical deliberation commence? Answer: another, different type of desire. The desire for the goal and the end, or as Aristotle puts it, "the that for the sake of which." Whereas *choice* is the desire for the means, *wish* is the desire for the goal and the end. And as we shall see, *wish* has unique causal powers insofar as it is the final cause of the craftsman's choices and subsequent actions. Moreover, a special kind of wish that I refer to as *calling*, is the *efficient* cause of the various "inner resources" that comprise the other components of the craftsman's motivational system.

Wish

Wish, according to Aristotle, is concerned with "the that for the sake of which" (*NE* 1113a) and the end. And because wish is concerned with the "end," it is conceptually connected to Aristotle's notion of the "good" of an action. Recall that Aristotle identifies the good of an action as its goal or end (*NE* 1097a). Wish is the *desire* for the goal and the end.

Because wish is desire for the goal and the end of action, it is a desire that has unique causal powers. First, it is the final cause of the actions that are performed for its sake. But recall that final causality "comes in degrees." Some goals are necessary conditions for the realization of other "more final" goals. Indeed, when we speak of the goal or an end of an action it is important to understand that, if the goal is sufficiently complex, its realization requires the realization of other mini-goals or "subroutines." We discussed this earlier in connection with hypothetical necessity: each step that is hypothetically necessary for the attainment of an end is a mini-goal whose realization is necessary for the realization of a most final goal. Unfortunately,

Aristotle uses the term wish (*boulêsis*) to refer to the desire for *any* goal, regardless of its finality. In other words, "wish" can refer to either the desire for a mini-goal *or* to the desire for the ultimate, most final goal. Since it is important to distinguish between these two desires, I will henceforth refer to the craftsman's desire for the ultimate, most final goal as the *calling*. Thus, *qua* pen-maker, Namiki's *calling* is the desire to make pens. Similarly, the *calling* of a doctor is the desire to produce health; the *calling* of the military strategist is the desire to achieve victory; and the *calling* of an architect is the desire to build a house. We will use the term *wish* to refer the desire for some goal or other. But, the *calling* is a special kind of wish. Let's first talk about the general notion of *wish*.

Wish and pleasure

Aristotle's conception of wish requires that we sever it from notions like "dream" and "hope." For Aristotle, most wishes are simply *desires for an instrumental goal*. When Jiminy Cricket sings "when you wish upon a star, makes no difference who you are," he is using a notion of wish that does *not* capture Aristotle's notion of this term. Jiminy alludes neither to a goal nor to actions undertaken for the sake of a goal.

Given that most wishes are simply desires for instrumental goals, the realization of one's wish *may or may not be* pleasure inducing. Namiki may feel a sense of relief having returned from his trek to the market, but this sense of relief is far from pleasure and gratification. Then again, Namiki may *love* going to the market. The important point is that the realization of the object of wish is *not necessarily* pleasure inducing, and if it is, the sense of pleasure is of a "different grade" than other pleasures that result from desire satisfaction like eating or drinking. Consider doing laundry, for instance. Doing laundry is an instrumental goal—we do not do it for its own sake. Therefore, we do not desire this activity for its own sake—we merely desire to do it because it is necessary for some further end. When we satisfy the desire (wish) to do laundry, we may feel good that it is "out of the way," but this grade of pleasure pales in comparison to other pleasurable experiences. The take-home point is that the realization of wish is not necessarily pleasurable, although it could be.

Let's now turn to an important species of wish—the *calling*. As mentioned above the craftsman's calling denotes not just his desire for *any* goal, but the desire for the ultimate, most final goal.

Calling

The craftsman's calling is the most final object of desire of the craftsman—his ultimate goal—and hence is "the why" for everything that the craftsman does. The calling is both the most final cause of the craftsman's actions and an efficient cause of the other components of the craftsman's motivational system. Indeed, it is the calling that causes the practical thinking that, as we have seen, is a necessary element in man-made production. Aristotle states that desire for the ultimate end stimulates practical deliberation (*DA* 433a). And as we know, practical deliberation, that is, strategic planning, causes choice. Ultimately, therefore, the object of the craftsman's calling is a powerful causal agent. It is the most final cause of the craftsman's goal-directed bodily motions and the *efficient* cause of the thinking, wishes and choices that enable the craftsman to realize his calling.

Calling and pleasure

Does the realization of the craftsman's most ultimate object of desire necessarily result in pleasure? When Namiki completes his pen, thereby realizing his calling, does he consequently feel pleasure? Answer: maybe. Remember, the discussion of the motivational system thus far has been narrow, focusing solely on the inner resources that enable the craftsman *qua* craftsman to engage in goal-directed activity. Considered within this narrow context, a calling may or may not be pleasure oriented. Namiki, for instance, could be a slave who loathes making pens but is forced to do so by a cruel master. In this case, Namiki would surely *not* feel pleasure upon making his pens and consequently realizing his "calling." With each completed pen, Namiki most likely feels disdain towards his master and possibly towards the whole pen-making industry. Nevertheless, Namiki could *love* making pens, whether he was ordered to do so or not. Even as a slave, he could take great joy in making his pens. He could look at each pen as an opportunity to exercise and improve upon his beloved craft and each time his cruel master demanded "another 200 units," Namiki would mutter silently, "hah! the joke is on him!" In this case, the realization of Namiki's calling would *indeed be* connected with pleasure. The point is, though, that considered in connection with the motivational system of the craftsman *qua* craftsman, a calling is not necessarily connected with pleasure. We have been using the term "calling" merely as a functional term that denotes the ultimate end of a specific craft or art—an ultimate desire for a goal about which the individual may or may not be enthused to realize.

Later on, however, we will see a context in which the calling *is* necessarily connected with pleasure. For Aristotle recognizes that the calling not only functions within the motivational system of the craftsman, but also functions within the motivational systems of human beings *in general*. As we will see later, Aristotle says that all human beings, *qua* human beings, have an ultimate object of desire and therefore a calling, namely, to flourish and to be happy. When we take the long view, and consider the calling within the motivational system of a human *qua* human, calling *is* necessarily connected with pleasure. We will discuss this pleasure-oriented sense of calling later. For now, however, think of the notion of calling in the narrower context of the motivational system of the craftsman. In this narrower context, calling corresponds simply to the craftsman's desire for his ultimate end, and hence is the most final cause of the craftsman's actions: "in medicine this is health, in strategy victory, in architecture a house, in any other sphere something else, and in every action and pursuit the end" (*NE* 1097a).

Let's summarize our discussion about the craftsman's motivational system. We have described both the intellectual and the desiderative components that are necessary for the goal-oriented action in which production consists. Among the intellectual components are *form* and *strategic planning*. Namiki understands how the different components of a pen are interrelated to enable the pen to perform its distinctive function of writing, and that knowledge enables him to devise a strategic plan or blueprint for its production. Then Namiki needs to *implement* this plan, thereby implicating a number of different desires. There are three different types of motivational forces or desires: (1) a desire for the means (*choice*); (2) a desire for some goal in the future (*wish*); and (3) the desire for the most ultimate, final goal (*calling*). The table below summarizes the essential components in the motivational system of the craftsman *qua* craftsman:

The motivational system of the craftsman

Intellectual components

	Definition	**Example**
Form	The knowledge of the essence of the product and of how the different components of the product contribute to its function.	Namiki's knowledge of the art of pen making.
Strategic planning	Thinking that proceeds from the form resulting in the sequence of hypothetical necessities for production.	"I need to make silver pen. I then need silver. I do not have silver, I need to go the market to buy silver."

Desiderative components

	Definition	**Example**	**Pleasure upon satisfaction ?**	**Type of cause?**
Choice	Desire that is caused by strategic planning that is itself the efficient cause of goal-directed bodily motion. Desire that is the result of deliberation.	Namiki's desire to buy silver and subsequent walk to the market.	No.	*Efficient* cause of goal directed bodily movements.
Wish	Desire for a future goal.	Namiki's desire to walk to the market to buy silver.	Doubtful. Most wishes are instrumental. Perhaps "relief" for getting this goal "out of the way."	*Final cause* of goal-directed bodily movements.
Calling	The most final wish. Desire for the ultimate goal, for the sake of which all other goals are realized.	Namiki's desire to make a pen.	Possibly.	*Most final cause* of goal directed bodily movements AND *efficient cause* of strategic planning, wish and choice.

Natural Production

Now that we understand the essential elements of Aristotle's theory of man-made production, we are now in a position to see how this theory extends to natural or biological production. Final causality and hypothetical necessity are not only involved in the production of man-made objects like pens, but also are involved in the production of *natural* objects such as plants, dogs, and human beings. Aristotle sees a profound similarity between craftsmen and "nature." Both are similar in that both are responsible for the orderly making of a thing. Just as pens do not "pop into existence" but instead result from Namiki's goal-oriented actions, so too, living organisms like human beings and dogs do not simply appear. Rather, they are the result of an orderly *natural* goal-oriented process.

Calling vs target

As we saw in the context of man-made production, the calling of the craftsman's actions is the desire for the completed product—for example, the silver pen. Keep in mind that a calling is necessarily connected with a motivational system. A calling is a type of *desire*. In natural production, however, production and biological development do not commence from a desire in a motivational system. Therefore, the notion of a "calling" is inapplicable to natural production. Nevertheless, there is an ultimate end or "goal" for the sake of which all prior developmental stages occur. I will use the term *target* to refer to the ultimate goal towards which development unfolds.

The distinction between the *calling* and the *target* corresponds to the two dimensions of final causality mentioned earlier. Recall that I distinguished the *objective dimension* of final causality from the *desiderative dimension*. The objective dimension corresponds to the operation of final causality in the objective, natural world. The desiderative dimension, on the other hand, corresponds to the operation of final causality within an individual's motivational system. The notion of target belongs to the objective dimension, whereas the notion of calling belongs to the desiderative dimension. Both the calling and the target are important components of Aristotle's view of flourishing, as we will see. To help the reader distinguish the concepts of *calling* and *target* the table below is provided:

Calling vs Target

	Definition	**Component of a motivational system ?**	**Essential to understanding Aristotle's view of flourishing?**
Calling	Desire for the ultimate goal. The cause of all subsequent desires.	Yes. A calling is a specific type of desire and is therefore a component of a motivational system. *Desiderative Dimension of Final Causality.*	Yes
Target	Ultimate goal of the natural biological development of a species.	No. Natural goal-directed processes do not proceed from desire. *Objective Dimension of Final Causality.*	Yes

In the following passages, Aristotle alludes to the target of development—the ultimate end or goal for the sake of which all prior developmental stages occur:

> There is "absolute" Necessity, which belongs to the eternal things, and there is "conditional" Necessity, which has to do with everything that is formed by the processes of Nature, as well as with products of Art, such as houses and so forth. If a house, or any other End, is to be realized, it is necessary that such and such material be available; one thing must first be formed, and set in motion, and then another thing; and so on continually in the same manner up to the End, which is the Final Cause, for the sake of which every one of those things is formed, and for which it exists. The things which are formed in Nature are in like case. (*PA* 639b–641a)

> *Because* man is such and such, *therefore*, the process of his formation must of necessity be such and such and take place in such a manner; which is why first this part is formed, then that. And thus similarly with all the things that are constructed by Nature. (*PA* 640b)

In these passages, Aristotle discusses the operation of final causality and hypothetical necessity in the realization of the target. Each step in biological development is for the sake of some further step, which, in turn, is for the sake of yet another further step and it is the ultimate end—the target—that emerges from this sequence.

Within nature, there is no man like Namiki who is behind the orderly natural process from which the target emerges. Nor does Aristotle invoke God's motivational system to explain the production of biological organisms like human beings and dogs. There is no desire and no motivational system directing this process. Rather, Aristotle says that the *soul* is behind the orderly natural process from which the development of a living being proceeds. A natural process culminates in its target because of soul.

The role of soul

Caution: Aristotle's notion of "soul" is radically different from the modern understanding of this term. For most of us, the word "soul" refers to something that is different from, and perhaps even independent of, the body. Further, few of us think of animals as having souls, and even fewer think that plants have souls. However, not only does Aristotle explicitly reject the view that the soul can exist apart from the body (*DA* 414a), but he also claims that plants and animals have souls. For Aristotle, anything that is living and develops has soul. In fact, "soul" for Aristotle is the final, formal, and efficient cause of the body:

> For the body, far from being one of the things said of a subject, stands rather itself as subject and is matter. It must then be the case that the soul is substance as the form of a natural body which potentially has life. (*DA* 412a)
>
> The soul, then, is the cause and principle of the living body, and as these are talked of in several ways, so is the soul the cause of the body in the three ways we have distinguished; for it is the cause as that from which the movement itself arises, and as that for whose sake it is, and as the formal substance of ensouled bodies. (*DA* 415b)

For Aristotle, the soul not only sets in motion the development of the body (efficient cause), but it also is the goal or end of its development (most final cause or target), and it is what the body is and *does* (formal cause) (*DA* 412a, 415b).

Obviously, Aristotle's account of soul is radically different from the conventional notion of soul. Aristotle conceives of the soul as a sort of craftsman that uses biological materials to make natural products. In the *De Anima,* he writes, "there is a parallel between a craft's need to use tools, and the soul's need to use the body" (*DA* 407b). For Aristotle, the human soul has more affinities to our modern notion of DNA than it does to some spiritual entity that can exist apart from the body. For Aristotle, the soul is the "natural engineer" of the body. Like our notion of DNA, soul directs the development of an organism.

Soul as form

In claiming that the soul is the form of a natural body, Aristotle is claiming that the soul contains the definition or essence of that body. Aristotle identifies the form of a body with its *nature* (*Phys.* 193a). Crucially, however, Aristotle believes that the form and nature of a particular organism is *not* something that exists at the outset. Rather, *form is something that develops and unfolds over time towards a specific end.* Again, natural products emerge from an orderly sequence of natural events just as man-made products emerge from an orderly sequence of the craftsman's actions.

As you may surmise, the idea that the soul is the "form" and "nature" of the body corresponds to the notion of "form" or "knowledge" that is involved in man-made production. Recall that Namiki's understanding of the form of pen enables him to devise the blueprint for its production. Form is information. When Aristotle says that the soul is the "form" of the body he is speaking of the informational system required for the development and growth of that body. But again, there is no "motivational system" that apprehends the form of an organism. Biological form is not in "God's mind" or in any specific motivational system. Rather, the form (i.e., soul) is intrinsic to the natural body that develops from it, in the same way that your particular genetic sequence or DNA is intrinsic to *your* body from which it developed. Again, for Aristotle the soul is not some spiritual entity that "lives on" after the body dies. Rather, the soul is akin to that which is responsible for the biochemical processes that begin the moment that the sperm fertilizes the egg. These biochemical processes just "go," without any motivational system driving them. Indeed, these natural processes follow the same internal logic as the development of man-made products. A specific order of steps is followed, and each earlier step is done for the sake of a further step, culminating in the end and completion of the organism. Thus, degrees of finality and serial goals are also part of the logic of natural production. Nature works like a craftsman, realizing a specific order of "mini-goals" that are necessary steps in the production of the organism:

> The existence and the formation of an eye is "for the sake of something." (*GA* 778a)

> Thus, those instrumental parts which are in their nature generative must always be there themselves prior to the rest, for they are *for the sake of* something else, as being a first principle. (*GA* 742b)

> [A]s soon as we begin to deal with those things that come into being through a process of formation, we find that there are several first principles—principles, however, of a different kind and not all of the same kind. (*GA* 743a)

Given that final causality and hypothetical necessity operate in natural production, an important question concerns the *end* of production. In the case of living organisms or species, what is the completed product, the goal, the target? What is the ultimate "that for the sake of which" biological development progresses? Is the target merely something alive? Or, is there some other stage of completion towards which nature strives? In the case of human beings, is the target an embryo? A fetus? An infant? An adolescent? An adult?

Crucially, Aristotle states that the target of natural production is realized when the *idion*—the characteristic function of that organism—is formed:

> It is not the fact that when an animal is formed at that same moment a human being, or a horse, or any other particular sort of animal is formed, because the end or completion is formed last of all, and that which is peculiar (*idion*) to each thing is the end of its process of formation. (*GA* 736b)

Aristotle is claiming that each species has something peculiar to it, some unique capacity. The formation of this unique capacity is the target of nature. Just as Namiki's calling is realized when he completes the silver pen, nature's target is realized when the *idion* of the species is formed. Immediately after the above remark from *Generation of Animals*, Aristotle states that the *idion* of a human being is reason or *nous*:

> That is why it is a very great puzzle to answer another question concerning Reason (*nous*). At what moment and in what manner, do those creatures which have this principle of Reason acquire their share in it, and where does it come from? (*GA* 736b)

In the case of the human soul, then, the target is reason. The various faculties, body parts, and the like possessed by man are, therefore, made for the sake of reason:

> In human beings, rational principle and mind are the end towards which nature strives. (*Pol.* 1334b)
>
> And as the body is prior in order of generation to the soul, so the irrational is prior to the rational. The proof is that anger and wishing and desire are implanted in children from their very birth, but reason and understanding are developed as they grow older. (*Pol.* 1334b)

Now that we understand the mechanisms behind both man-made and natural production, we are in a position to return to the function argument. As we will see, Aristotle's views about production show that the moves he makes in the function argument are defensible indeed.

The Function Argument Revisited

Earlier, we noted a number of alleged difficulties with Aristotle's function argument. First, Aristotle makes a number of strange analogical arguments. He analogizes human beings to artists/craftsmen—arguing that since the flourishing of an artist resides in the artist's function, it therefore follows that the flourishing of man resides in man's function.

Second, Aristotle seems to commit the fallacy of composition when he concludes that man has a function from the fact that body parts have functions. But again, the fact that the parts of an entity have a certain property does not imply that the entity as a whole has the property. Each sentence in Tolstoy's mammoth 1300 page book *War and Peace* is relatively short. However, the book itself is not short.

Third, Aristotle argues that the function of man can be discovered if one discovers what is "*idion*" or peculiar to man. Despite the numerous things that are peculiar to man (giving to charity, taxidermy, etc.), Aristotle ignores all of these candidates and selects, without explanation, rational activity.

Fourth, there is the further fact that the conclusion of the function argument is ambiguous, and therefore it isn't at all clear *what* Aristotle's view of flourishing is. The conclusion of the function argument, recall, defines flourishing as "excellent activity of the rational element." But Aristotle acknowledges several different kinds of excellent rational activity: ethically virtuous activity (which includes the virtues of character and the activity of practical wisdom) and contemplative activity. The conclusion of the function argument *does* indicate that flourishing is excellent activity of the rational element that is best and *teleiotatos*. But which activity is best and *teleiotatos*—ethically virtuous activity, contemplative activity, or both? As we know, intellectualists and inclusivists disagree on the answer to this question and therefore propose rival interpretations of Aristotle's conception of flourishing.

Let's first address Aristotle's seemingly problematic analogical arguments. The analogies that Aristotle draws between specific tradesmen and *homo sapiens* are quite natural given his views about the motivational system of the craftsman *qua* craftsman, on the one hand, and that of the human *qua* human, on the other. For he believes that both are driven by an ultimate object of desire, that is, a calling.

Considered within the context of the motivational system of the craftsman *qua* craftsman, a calling is the desire for the completed product. The calling, recall, refers to the ultimate desire—the most final cause of all subsequent wishes, choices, and actions of the craftsman *qua* craftsman. Everything Namiki does *qua* pen maker: from the buying of the materials, to the molding, shaping, and so on, is done for the sake of making a pen. But let's consider Namiki not as a pen-maker, but as a human being. Stripped of his professional role as pen-maker, Namiki is, after all, a human being. And *qua* human being, according to Aristotle, everything Namiki does, chooses, wishes, and so on, stems from one ultimate desire—his desire for *eudaimonia* or flourishing. Thus, flourishing is the calling of a human being—it is the ultimate goal for the sake of which all subsequent "mini-goals" are pursued. And when we consider this fact, the move in the function argument in which Aristotle draws an analogy between craftsmen and human beings is defensible indeed. Let me explain further.

The Human Being's Motivational System

In Book I of the *NE,* Aristotle informs us that the motivational system of a human being *qua* human being is driven by a desire for ultimate goal—a calling—which is the most final cause of all of the wishes, choices, and subsequent actions of the individual. As Aristotle notes, if the motivational system of a human *qua* human was *not* driven by one ultimate object of desire, our desires would be "empty and vain":

> Every art and every inquiry and similarly every action and pursuit, is thought to aim at some good; and for this reason the good has rightly been declared to be that at which all things aim . . . If, then, there is some end of the things we do, which we desire for its own sake (everything else being desired for the sake of this), and if we do not choose everything for the sake of something else (for at that rate the process would go in to infinity, so that our desire would be empty and vain), clearly this must be the good and the chief good. (*NE* 1094a)

Aristotle's point is simple: there must be one ultimate goal—*the* desire which is the cause of all our other desires. Consider someone who desires an education, not as an ultimate end, but as a means of obtaining a job. She then *chooses* to become educated for the sake of getting a job. Further suppose that she chooses to get a job not for its own sake, but because it is a means to satisfy some further desire, namely, her desire for wealth. Ultimately, according to Aristotle, this chain of desires for the "mini-goals" must stop at some *most* final desire for whose sake we desire everything. This ultimate desire is our desire for *eudaimonia* or flourishing, and thus flourishing is the calling of a human *qua* human. Because an ultimate object of desire—a calling—drives the actions of both craftsmen and human beings, we can say that both have similar motivational systems.

Both craftsmen and human beings are *teleological systems* insofar as desires for ultimate ends or goals drive their respective motivational systems. However, recall Sartre's objection to the analogy between craftsmen and human beings mentioned earlier. Sartre would object to this analogy on the grounds that, whereas the content of the craftsman *qua* craftsman's calling is determined by his craft, the *content* of the human *qua* human's calling is not. Put differently, the calling of the craftsman *qua* craftsman has a specific, prescribed content that is not "up to" the craftsman. "In medicine this is health, in strategy victory, in architecture a house, in any other sphere something else, and in every action and pursuit the end" (*NE* 1097a). However, the calling of a human *qua* human has *no* particular content. The content is "up to" the individual, says Sartre. Existence, as he says, precedes essence.[3] A human being determines the content of her calling, and thus she is most *un*like a craftsman.

Sartre is partly right and partly wrong. First, note that the disagreement between Sartre and Aristotle concerns the appropriate degree of conceptual abstraction. At one level, the content of a human *qua* human's calling varies from person to person. But Aristotle acknowledges this fact. Indeed, he notes that the *content* of one's calling varies from person to person: some say it is pleasure, some say it is honor, some say it is wealth (*NE* 1095a). Aristotle's distinction between the "real" and the "apparent good" (*NE* 1113a) implies that content of the human *qua* human's calling is, in one sense, "particular" and "subjective." This individual variance notwithstanding, we can nevertheless make a universal claim about the *structure* of a human being's motivational system, namely, that there is one single desire that is the ultimate end of *all* other desires. This universal claim is simply about the "hardware" of the human being's motivational system. The claim is that various components of this system—wishes, choices, practical thinking, and so on,—are elements which emerge from one single "parent" desire, the calling, for which they exist to serve. At this very general level of abstraction there is a profound similarity between the motivational system of the craftsman *qua* craftsman and the human *qua* human. Just as the intellectual and desiderative components of the craftsman *qua* craftsman's motivational system emerge from his calling to produce a pen and exist to serve that calling, so too the intellectual and desiderative components of the human *qua* human's motivational system emerge from his calling for *eudaimonia* and exist to serve that calling.

Aristotle, however, believes that some people can get it wrong (*NE* 1113a). Those for whom the calling is bodily pleasure, honor, or wealth get it wrong (*NE* 1097b). And it is here where he may very well depart from Sartre. Indeed, the conclusion of the function argument stands for the proposition that the *idion* of the human being determines the correct content of this species' calling. For Aristotle, *the content of a human being's calling is prescribed by human nature.* This is precisely why those who identify their calling as bodily pleasure, honor or wealth get it wrong: their actions are not proper given their nature. Now whether Aristotle is correct in claiming that the content of a human being's calling as prescribed is *one* question, and Sartre may object to Aristotle on this claim. However, given *Aristotle's* belief that the content of the human being's calling *is* prescribed, then his analogies between craftsmen and human beings make *perfect* sense. First, both have motivational systems that are driven by an ultimate desire—a calling. Furthermore, for both craftsmen and human beings, the content of their respective callings is *prescribed.* The content of the doctor's calling is not "subjective" and therefore is not "up to" a doctor to determine the content of her calling *qua* doctor. Nor is the content of the human *qua* human's calling subjective. It is not "up to" a human to determine the content of her calling *qua* human. Given the deep similarities between the motivational systems of craftsmen and human beings, Aristotle's analogies between the two make good sense indeed. The following table summarizes these similarities:

	Driven by an ultimate object of desire?	**Is the content of the calling prescribed?**
Motivational system of craftsman *qua* craftsman	Yes: Calling: desire to realize the end of the craft. Calling is the efficient and most final cause of all the other intellectual and desiderative components of the motivational system.	Yes. The craft determines the content of craftsman's calling. The content of the craftsman's calling is not "subjective." It is not "up to" a doctor to determine the content of his calling *qua* doctor.
Motivational system of human *qua* human	Yes: Calling: desire for *eudaimonia* or flourishing. Calling is the efficient and most final cause of all the other intellectual and desiderative components of the motivational system.	Yes. The *idion,* nature's target, determines the content of human *qua* human's calling. The content of man's calling is not "subjective." It is not "up to" man to determine the content of his calling *qua* human being.

Of course, Sartre and many others would vehemently disagree with Aristotle's claim that the content of man's calling is prescribed. We will discuss this objection in more detail later on. For now, however, let us turn to the criticism that Aristotle commits the fallacy of composition in the function argument. As we will see, once one considers Aristotle's beliefs about natural production, this objection becomes groundless.

Recall that Aristotle reasons, seemingly fallaciously, from the fact that each of the body parts has an *ergon,* to the claim that man has an *ergon.* However, when we keep in mind that Aristotle believes that the parts of the human body do not exist for their own sake, but exist for the sake of some other end, namely, the *idion* and the target, this inference makes perfect sense. Aristotle believes that nature designs each organism to perform a specific function and this function is determined by the organism's *idion.* Thus, Aristotle's inference that man has an *ergon* from the premise that the eye, hand, and foot each have an *ergon* is not fallacious. Rather, it is an expression of Aristotle's belief that the parts of the body are there for the sake of nature's target—the *idion* of the organism.

Given Aristotle's views about both man-made and natural production, then, we can see that the analogies that he makes in the function argument make complete sense. First, human beings are like craftsmen insofar as both possess motivational systems that are driven by an ultimate object of desire, that is, a calling. Moreover, human beings are like craftsmen insofar as the content of their ultimate end is prescribed. In the case of the craftsman, the content of his calling is prescribed by the nature of his craft. In the case of a human being, the target of man's natural development—that is, the *idion*, prescribes the content of man's calling.

Further, Aristotle commits no compositional fallacy when he infers that man has an *ergon* from the fact that the body parts of man each have their own *ergon*. For in the case of human beings, the body as a whole exists for some rational activity. As Aristotle says, it is because he has intelligence (*phronêsis*) that man has hands:

> Hands are an instrument; and Nature, like a sensible human being, always assigns an organ to the animal that can use it . . . it is not true to say that man is the most intelligent animal because he possesses hands, but he has hands because he is the most intelligent animal. (*PA* 687a)

The doctrine that there is a target of natural development acquits Aristotle from the charge that he commits the fallacy of composition.

The above remarks also help us to understand why Aristotle selects rational activity as the *idion* of man, rather than, for example, prostitution or ant burning. The human body, including its hands, tongue (*PA* 660a), lips (*PA* 659a), eyebrows and eyelashes (*PA* 658b), larynx (*PA* 664a), and so on, exists for the sake of nature's target. Indeed, the production of each of these parts is a "mini-goal" that is "set" for the sake of nature's target. Aristotle selects reason as the *idion* because reason is the target of human development (*GA* 736b; *Pol.* 1334b). The function argument claims that man's *ergon* can be discovered if we can determine what is *idion* to human beings. And now we understand that this means that man's *ergon* is known when one understands the target of human development.

Now, being mindful of Aristotle's views about man-made production, nature, and biological development may shed light on some of the premises that he makes in the function argument, but what does it show about the meaning of "best and *teleiotatos*" in the conclusion of this argument? Does an appeal to Aristotle's biological works help us to reconcile the debate between intellectualists and inclusivists? Which is best and *teleiotatos*, contemplative activity (as intellectualists insist) or a *combination* of contemplative and ethically virtuous activity (as urged by inclusivists)?

Aristotle's views about man-made and natural production not only help us to understand the function argument, but they also illuminate the nature of the relationship between ethically virtuous and contemplative activity. We have noted that there are two dimensions at which final causality operates in Aristotle's system: the objective dimension which concerns the operations of natural, biological development, and the desiderative dimension which concerns the goal-oriented actions of human beings. The *target* lives in the objective dimension, whereas the *calling* lives in the desiderative dimension. I have argued that the target determines the content of the calling that should drive a human being's motivational system.

The calling of a human being is rational activity. Flourishing, on the other hand, consists in the perfection of this activity. To flourish is to perform man's calling with excellence. This includes contemplative activity. But one's motivational system (and hence the desiderative dimension of final causality) must be *cultivated* and *developed* in order to engage in excellent contemplative activity. One *becomes* a *sophos*, and not just any motivational system has the potential to carry its possessor towards contemplative excellence.

In the following chapter, I argue that only a motivational system that possesses ethical virtue has the potential to lead its possessor towards contemplative excellence. As I argue, ethical virtue is necessary for the development of contemplative activity for two reasons. First, it ensures that the motivational system is driven by a desire for the activities that lead to contemplative excellence. Second, it implies an excellence of belief, which, given Aristotle's epistemology, is necessary for knowledge and therefore for contemplative excellence. It is to these arguments that we now turn.

Chapter 3

Ethical Virtue as a Developmental Prerequisite for Contemplative Excellence

Introduction

As we know from previous discussion, one of the difficulties surrounding the interpretation of the function argument concerns Aristotle's remarks in Book X of the *NE*. There, Aristotle seems to exalt contemplative activity as the highest and best activity, and to regard a life in accordance with ethical virtue—"the other kind of virtue"—as happy in a "secondary" sense (*NE* 1178a–1179a). Both intellectualists and inclusivists agree that contemplative activity is an essential ingredient in a flourishing life. What they *disagree* on, however, is whether it is the *only* such ingredient. Intellectualists say that it is, inclusivists say that it isn't. According to inclusivists, both contemplative *and* ethically virtuous activity are essential components of a flourishing life.

Ethically virtuous activity is a complex rational activity that includes both the virtues of character and practical wisdom or *phronêsis*. Aristotle regards the virtues of character and *phronêsis* as excellences of different capacities of the soul (*NE* 1138b–1139a). I agree with the inclusivist position that ethical virtue is an essential component of flourishing. However, inclusivists have failed to recognize that ethical virtue is a *developmental prerequisite* of excellent contemplative activity.[1] In order to see how ethical virtue is necessary for excellent contemplative activity, we need a clear understanding of both of its components. As I will show, the virtues of character and *phronêsis* imply motivational and epistemological excellences that are necessary for contemplative excellence.

There are three stages to my argument. Stage one focuses on the part of the soul involved in the virtues of character, namely the desiderative faculty or appetite. I refer to this part of the soul as the "affective" part of the soul and refer to its excellence as "affective virtue." Contemporary psychologists use the term "affect" to refer to the dimension of human experience concerned with pleasure and pain and I use that terminology here.

The affective part of the soul is extremely important because it is necessarily involved in action (*DA* 433a). I argue that the affective part of the soul plays the *commanding* role in a human being's motivational system. It is ultimately affect that drives activity. But crucially, affect can be *wrong* and so human beings can be driven to pursue activities that do not conduce to human flourishing. Understanding this point is essential to understanding the nature of the virtues of character or "affective virtue." As we will see, acquiring affective virtue is tantamount to refining a human being's motivational system so that is commanded by the correct affect.

After focusing on the affective part of the soul, I move to stage two of the argument and shift focus to the other component of ethical virtue, namely, *phronêsis*. Aristotle regards *phronêsis* as a *rational* or *intellectual* excellence of the belief or opinion-making faculty (*NE* 1140b). As we shall see, *phronêsis* implies a number of epistemological excellences that are necessary for the acquisition of knowledge, among them, true belief. This fact is important for appreciating how *phronêsis* is developmentally necessary for excellent contemplative activity. For true belief is the starting point of scientific knowledge or *epistêmê*.

The discussion of affective virtue and *phronêsis* will demonstrate that ethical virtue is necessary for contemplative excellence. Contemplative excellence is a master ability that requires both epistemological *and* motivational preconditions. Ethical virtue not only provides its possessor with true belief, but it also provides its possessor with the *intrinsic motivation* to engage in the continuous intellectual activity that is necessary for contemplative excellence. Put simply, ethical virtue provides the epistemological and motivational starting points for contemplative excellence. Because he possesses a specific motivational system, the ethically virtuous individual is poised to engage in the activities through which one becomes *sophos* or wise.

This brings me to the third stage of my argument which is concerned with the students of the *NE*. The *NE* is a transformative tool, written for a very specific kind of student: one who is mature and is not governed by emotions and is the beneficiary of a good upbringing (*NE* 1095a–b). Further, the purpose of the *NE* is not to *know* what flourishing is, but to flourish—to "become good" (*NE* 1103b). The end is not knowledge but action (*NE* 1095a). I argue that Aristotle requires a specific kind of student because his aim is to further refine the motivational system of his pupil. In my view, Aristotle's aim is to further cultivate his apprentice so that his actions more accurately conduce to flourishing. Just as a craftsman uses a finishing tool to perfect his product, Aristotle uses the *NE* to refine the student's calling—the ultimate object of desire that drives the student's actions. To achieve this aim, Aristotle helps the student to further develop the epistemological and motivational starting points for contemplative excellence. By focusing on how the *NE* affects the motivational system of its student, we will see that Aristotle intended to promote the moral development of his students by engaging them in an activity aimed at putting them on the path to contemplative excellence.

With this preliminary roadmap out of the way, let's begin by discussing the affective part of the human soul.

The Nature and Development of Affective Virtue

In our earlier discussion about the soul, we noted that Aristotle conceives of the soul as something functionally similar to DNA rather than as a "spiritual" entity that survives bodily death. The soul, recall, is the "final cause" of the body and hence acts like a craftsman insofar as it directs the natural development of the organism (*DA* 415b).

In addition to describing the soul as the final cause that guides and directs biological development, Aristotle sometimes describes the soul *qua* developed and hence as the *set of capacities* possessed by a mature organism. In the *De Anima,* for instance, Aristotle discusses a variety of capacities that are possessed by living organisms: the nutritive, the perceptive, the desiderative, the locomotive, and the rational (*DA* 414a–415a). Not every living organism possesses *all* of these capacities, however. For example, plants have only the nutritive capacity and do not have the capacity for locomotion. Horses and dogs, however, have the nutritive, desiderative, *and* the locomotive capacities. Man, a unique animal, not only possesses the nutritive, desiderative, and locomotive capacities, but also possesses the capacity for rational activity (*DA* 415a).

The uniqueness of man's desiderative faculty

Because man is capable of rational activity, the capacity to reason imparts a distinctive trait on to man's desiderative or appetitive faculty. The desiderative faculty is an extremely important capacity for various reasons, chief among them is its connection to locomotion, movement, and action (*DA* 433a). In fact, Aristotle claims that the desiderative faculty is the *principal cause* of bodily movement (*DA* 433a).

Man's desiderative faculty is unique because it has the ability to obey *or* disobey reason:

> [There is] also another natural element beside the rational principle, which fights against and resists that principle. For exactly as paralysed limbs, when we intend to move them to the right, turn on the contrary to the left, so is it with the soul; the impulses of incontinent people move in contrary directions. But while in the body we see that which moves astray, in the soul we do not. No doubt, however, we must none the less suppose that in the soul too there is something contrary to the rational principle, resisting and opposing it . . . but the appetitive and in general the desiring element in a sense shares in it, in so far as it listens to and obeys it . . . That the irrational element is in some sense persuaded by a rational principle is indicated also by the giving of advice and by all reproof and exhortation. (*NE* 1102b–1103a)

According to Aristotle, the desiderative faculty can fight and resist the rational principle, *or* it can listen to and obey it. But what precisely does he mean here? Specifically, what does Aristotle mean by "rational principle"? For Aristotle claims that there are two parts of the soul that can grasp a rational principle: the "scientific" and the "calculative":

> [L]et it be assumed that there are two parts which grasp a rational principle—one by which we contemplate the kind of things whose originative causes are invariable, and one by which we contemplate variable things . . . Let one of these parts be called the scientific and the other the calculative; for to deliberate and to calculate are the same thing, but no one deliberates about the invariable. Therefore the calculative is one part of the faculty which grasps a rational principle. (*NE* 1139a)

Remember, Aristotle conceives the term "reason" *broadly*. "Rational activity" not only encompasses the highly cerebral activity of contemplation, but also includes practical deliberation or "strategic planning" connected with goal-directed action. Thus, the "rational part" of the soul has two parts: the "scientific" which is earmarked for contemplative reason, and the "calculative" which is earmarked for practical reason or deliberation. Aristotle unequivocally states that it is the calculative part or practical reason that is concerned with goal-directed action (*NE* 1139b). He reiterates this point in the *De Anima* when he says that contemplative reason "never thinks what is practicable" and "never says anything about an object to be pursued or avoided" (*DA* 432b). Practical reason, however, is concerned with pursuing the good (i.e., goal) that is realized through action (*DA* 433a–433b). Thus, the "rational principle" which man's desiderative faculty may or may not obey is grasped by *practical* reason or the calculative part of the rational soul.

The uniqueness of man's actions

The fact that man's desiderative faculty is unique implies that man's *actions* can be unique. The desiderative faculty controls movement and locomotion (*DA* 433a). Thus when desire listens to practical reason, man engages in a special kind of movement and locomotion.

This special kind of movement is not voluntary movement *simpliciter*, however. As Aristotle says, animals can engage in voluntary movement (*NE* 1111b), but man is capable of more than mere voluntary movement.

The special kind of movement in which man can engage is a movement directed by a rational principle or a "rule" that is grasped by practical reason. Thus, man's desiderative faculty, and therewith his actions, can be directed by practical reason. How is movement informed by practical reason different from mere voluntary activity? We will pursue this question in more detail when we discuss affective virtue, but for now, we can note that through practical reason, a "rational principle" or rule can direct the body. The rational principle *pushes* the body to perform actions for the sake of some future goal as well as *pulls* or *restrains* the body from performing activities that interfere with the attainment of this goal.

Consider, for example, the bodily movement of the craftsman. The craftsman's actions proceed from a desiderative faculty informed by a strategic plan or blueprint. In a significant sense, it is the plan or the blueprint that pushes the craftsman's body to perform certain actions. And, of course, to "stay on target" and realize his end, the craftsman's desiderative faculty must also *restrain* the body from performing actions that undermine the achievement of his end. For example, Namiki would not succeed in making his silver pen if he acted on his desire to hoard the silver and use it to feed his gambling addiction. The realization of his goal requires both forward movement and self-restraint. By letting the strategic plan direct his body, Namiki listens to a rule or "rational principle."

In the goal-directed actions of the craftsman, then, practical reason, via a strategic plan, guides the craftsman's body by guiding the desiderative faculty, pushing it and restraining it where necessary. This dynamic *may* be true when we consider the actions of the human *qua* human. In order words, a strategic plan *may* guide the actions of the human being. I say "may" because human beings are not born with a desiderative faculty that is naturally guided by a strategic plan. And human beings are certainly not born with a desiderative faculty that is naturally guided by the *correct* strategic plan. The actions of children, for example, are impulsive and spontaneous and are not guided by practical reason (*NE* 1111b). Indeed, the desiderative faculty of infants operates with no restraints, thereby causing the unique behavior of the "immature" (*Pol.* 1260a).

Human beings possess the *potential* for acting pursuant to a rule rather than passion. As we shall see later, Aristotle believes that the actualization of this potential can only come about through a certain kind of training. But when this training is complete, human beings engage in a type of action that is *uniquely human.* But again, it is not merely voluntary or "self initiated" action that is uniquely human, for animals, too, act voluntarily (*NE* 1111b). It is voluntary action *that is directed by a strategic plan apprehended by practical reason* that is unique to man. Ideally, of course, man's actions should be guided by the *correct* strategic plan. Man has the potential for acting in this way, but the potential may fail to be realized. Human action may be "mere voluntary" action which is common to both animals and immature human beings. But human action may also be *reason directed* voluntary action which is a mature, trained type of action that is unique to man.

Affect and the desiderative components

We noted earlier that pleasure and pain are irrelevant to the motivational system of the craftsman *qua* craftsman. Namiki, recall, need not feel pleasure when realizing his calling *qua* pen-maker. But things are different when we consider the motivational system of the human *qua* human. Where human action and the calling of the human *qua* human are concerned, affect is indeed relevant and profoundly so. In fact, affect is necessarily connected to the desiderative faculty (*DA* 413b).

When we discussed the motivational system of the craftsman, we noted that desire can function as both an *efficient* cause of action *and* as a *final* cause. And we further noted that desire can function as a *most final* cause of the craftsman's actions. There are three different types of desires that comprise the desiderative component of the craftsman's motivational system, all of which function as different types of causes of the craftsman's bodily movements: (1) *choice* which is the efficient cause of these movements; (2) *wish* which the final cause of these movements; and (3) *calling* which is the *most* final cause of the craftsman's movements.

As we shall see, the fundamentals of this story remain the same when we consider the motivational system of the human *qua* human. However, because the desiderative component of a human being's motivational system is *necessarily* connected to affect, affect is therefore necessarily connected to human action. Because affect is necessarily connected to the desiderative faculty (*DA* 413b), and because desire can be either an efficient or a final cause of action, then affect can be either an efficient or a final cause of action.

Further, affect can also function as an effect/response to some stimulus. Consider anger. Anger is clearly a painful emotion that can function as an effect or response to someone's behavior. Aristotle seems to speak of anger in this way when he speaks about being angry at the right things and with the right people (*NE* 1125b). But anger can also function as a *cause* of further behavior. In fact, affect can function both as an efficient and as a final cause of action.

Affect as an efficient cause of action

In order to see how affect can function as an efficient cause of action let's stick with our example of anger. Suppose that your beloved forgets to call, thereby causing you to feel angry. In this case, you experience anger as a response—an *effect*—of your beloved's failure to call. But of course, your anger could, in turn, cause you to drink, eat, and/or gamble excessively. In this case, your painful emotion of anger is a motivator—an efficient cause—of further behavior. Parallel cases can, of course, be devised for pleasurable emotions. For instance, your beloved may surprise you with a thoughtful note that is secretly mailed to you at work, thereby causing you to feel the pleasing emotions of joy, elation, and gratitude. Thus, these pleasurable emotions are experienced as an effect/response to your beloved's actions. These pleasurable emotions can, in turn, cause you to smile and phone your beloved to express thanks. If so, your pleasurable affect functions as an efficient cause of your behavior. Both pleasurable and painful affect can function as efficient causes of behavior.

Affect as a final cause of action

In addition to functioning as an efficient cause of one's actions, pleasurable and painful affect can also function as *final* causes of action. Importantly, pleasure and pain are not only experienced in connection with our reactions and actions towards others, but are also experienced in connection with *activities* (*NE* 1175a). Pleasurable activities motivate us to pursue them, whereas painful activities cause us to avoid them (*NE* 1175b). An individual who finds music pleasurable pursues musical activities (*NE* 1175b), whereas an individual who finds arithmetic painful avoids arithmetic (*NE* 1175b). Aristotle also claims that if an individual is faced with two pleasurable activities, he pursues the activity that is more pleasurable. In his amusing example, Aristotle mentions an individual who loves flute playing more than discussion. Such an individual is unable to focus on the discussion at hand because his attention is enraptured by the mellifluous sounds of the flutes providing the ambient music for the discussion (*NE* 1175b).

Pleasurable affect can function as a final cause of action when it is connected to an object or activity that is a future goal for an individual. For instance, suppose I find visiting New York City pleasurable, and consequently organize a trip to visit this summer. In this case, an activity that causes pleasant affect (visiting New York) functions as a future goal and hence as a final cause for all of the antecedent and instrumental goals (booking a flight, getting a hotel room) that are necessary for its realization. Painful affect can also function as a final cause of action. If one anticipates an activity that causes pain, one may engage in a sequence of actions designed for the sake of avoiding this activity. For instance, if I find talking to a specific individual painful, I may choose to avoid places at which this individual will be present. Thus, pleasurable and painful affect can function as final causes of action.[2]

Affect as the most final cause of action

As Aristotle tells us, all of a human being's desires and actions proceed from an ultimate desire—the desire for *eudaimonia* or flourishing (*NE* 1094a). It is the desire for flourishing which provides the ultimate and most final "that for the sake of which" of all of our desires and actions (*NE* 1094a). Without this desire all other desires would be "empty and vain." Thus, the desire to flourish plays the *commanding* role in a human being's motivational system. We have referred to this commanding desire as the *calling* of the motivational system. And given that desire is for the pleasant (*DMA* 701b), then one's calling is also for the pleasant. Accordingly, Aristotle notes that the fact that all men pursue pleasure indicates that pleasure is connected with the most final cause of their actions (*NE* 1153b). In Aristotle's view, the calling of the human *qua* human is necessarily connected to pleasure.

To sum up our discussion, then, in the motivational system of a human being, affect can function as an efficient cause of action, as a final cause of action, or as a most final cause of action. These three causal roles correspond to the three causal roles of desire: some desires are the efficient causes of action, some are final causes, and one—the calling—is the most final cause of action.

The primacy of affect

When we consider the relationship between pleasure and desire it is important to note that pleasure is primary. One ultimately desires some activity *because* it leads to pleasure. Thus it is one's sense of pleasure that determines what the individual chooses and does. It is almost as if one's sense of pleasure is a key of some sort—that "scans" possible ends and settles on those with which it best fits.

One might object to the claim that pleasure determines what the individual chooses and does by pointing out that we choose to do many things even though they are *not* pleasurable. For example, I plan to take the Michigan Bar exam. But I do not desire to do this because I believe that I will get pleasure from taking the Bar exam (although I may). Rather, this goal is purely "instrumental"—it is merely a means to a further goal, that is, practicing law. So, doesn't this refute the claim that pleasure determines what the individual chooses and does?

Actually, this argument overlooks the distinction made earlier between desire *qua* choice and desire *qua* final cause. A choice, recall, is a desire to bring about the conditions which are necessary to realize some further end. Granted, the realization of choice is not necessarily pleasure inducing. In order for me to eat, I must choose to go food shopping, but when I engage in this act it is not pleasure yielding. To return to the Bar exam example, I must take the Bar exam, but when I engage in this act it is not pleasure inducing. Yet, it is precisely *because* I believe that practicing law will bring me pleasure that I choose to take the Bar exam. Indeed, to realize my goal of practicing law I am required to do *lots* of things that I find displeasing (taking the LSAT, going to law school, taking the Bar exam, etc.). But it is my ultimate end of practicing law, and the pleasure that I anticipate from its attainment, that causes me to perform these unpleasant actions. Pleasurable affect, therefore, is the final cause of the unpleasant actions that I perform. That Aristotle agrees with this claim is evidenced when he discusses the motivation of a boxer. For although the boxer endures pain, he does so for the sake of the pleasure that he gains from the honor that comes with victory (*NE* 1117b).

In order to see that pleasurable affect commands the motivational system of the human *qua* human, one needs to take the "long view." And when one understands the most final cause(s) as to *why* one desires to do something that is painful, the answer will necessarily involve some object or activity that the individual finds pleasurable. Pleasurable affect plays the commanding role in a human being's motivational system, and hence determines the content of its calling. Pleasurable affect is the most final cause of a human being's actions.

The particularity of affect

Importantly, although all human action proceeds from a calling, the specific pleasurable affect that determines the content of this desire varies from person to person. Human beings possess the same "motivational hardware," that is, a motivational system driven by the desire to flourish, but the "software"—the pleasurable affect—that drives each *particular* system varies from person to person. Thus, the hedonist is motivated to pursue the pleasures of the body (*NE* 1153b, *NE* 1176b), the geometer the pleasures of geometry (*NE* 1175a), and the lover of flute music the pleasures of flute music (*NE* 1175b). People for whom bodily pleasures dominate are motivated to pursue one class of activities, and people for whom rational pleasures dominate are motivated to pursue a different class of activities:

> [I]n the case of men at least; the same things delight some people and pain others, and are painful and odious to some, and pleasant to and liked by others. (*NE* 1176a)

Indeed, history is abound with examples of different people being driven to pursue different pleasurable activities. Cole Porter, the famous American composer who came from a family of attorneys and who attended Harvard Law School for just a year, was moved not by the pleasures of law, but by the pleasures of music. W. H. Murray, a Scottish mountain climber who advanced the sport, was moved not by the pleasures of music, but the pleasures of climbing.

Correct affect

Now, Aristotle believes that the pleasurable affect that drives human action can be *wrong*. His discussion of the "good" and the "apparent" good (*NE* 1113a, *DA* 433a) expresses his belief that the ultimate "that for the sake of which" for which we do everything can have the *incorrect affective content*:

> That wish is for the end has already been stated; some think it is for the good, others for the apparent good. Now those who say that the good is the object of wish must admit in consequence that that which the man who does not choose aright wishes for is not an object of wish (for if it is to be so, it must also be good; but it may well have been bad); while those who say the apparent good is the object of wish must admit that there is no natural object of wish, but only what seems good to each man. Now different things appear good to different people, and, if it so happens, even contrary things. (*NE* 1113a)

Aristotle uses the word "wish" to refer to the ultimate desire that drives a human being's motivational system (*DA* 433a). Again, we have referred to this ultimate desire as the *calling* and we have said that pleasurable affect determines the content of the calling. Aristotle's point above is that the "good," that is, the pleasurable affect that comprises one's calling can be *wrong*. Indeed, he explicitly states that this error is due to pleasure and pain and hence affect (*NE* 1113b). The "real" or "true" calling refers to the specific affect that *should* be driving our motivational system. The "apparent" calling, however, refers to *incorrect* affect that can drive the motivational system.

The fact that our actions can proceed from the wrong affect is absolutely essential for understanding Aristotle's theory of affective virtue. As Aristotle states, the "whole concern" of the virtues of character is with pleasure and pain:

> Moral excellence is concerned with pleasures and pains; it is on account of the pleasure that we do bad things, and on account of the pain that we abstain from noble ones. Hence we ought to have been brought up in a particular way from our very youth, as Plato says, so as both to delight in and to be pained by the things that we ought. (*NE* 1104b)
>
> [I]f the virtues are concerned with actions and passions, and every passion and every action is accompanied by pleasure and pain, for this reason also virtue will be concerned with pleasures and pains. (*NE* 1104b)
>
> [T]he whole concern both of virtue and of political science is with pleasures and pains. (*NE* 1105a)
>
> [T]o enjoy the things we ought and to hate the things we ought has the greatest bearing on virtue of character. (*NE* 1172a)

One who possesses affective virtue pursues correct pleasures and pains and hence pursues the correct activities. Of course, the natural question raised by the notion of affective virtue is, in virtue of *what* is pleasure and pain correct? What is the standard in virtue of which one determines whether one is motivated by the correct pleasures? Aristotle acknowledges a smorgasbord of pleasurable activities available to human beings. There are pleasures that presuppose some sort of pain, such as the pleasure that we feel when drinking to satisfy thirst (*NE* 1173b). There are necessary pleasures of the body such as those experienced from sexual activity (*NE* 1147b). There are pleasures that we get from our senses (*Met.* 980a). There are also rational pleasures such as learning and contemplative activity (*NE* 1153a). So the question remains, how does one determine which kind of pleasure provides the correct affective content for man's calling?

Affect and the target of nature

As we saw earlier, the function argument is central to Aristotle's account of flourishing and resolves the debate about the best life for a human being. But the function argument is not only an argument about the proper work and *idion* of man, but is also an argument about man's *proper pleasure*. Aristotle selects "rational activity" as the function of man not only because it is unique to human beings, but also because the pleasure derived from rational activity is the highest and best pleasure:

> Each animal is thought to have a proper pleasure, as it has a proper function; viz., that which corresponds to its activity. (*NE* 1176a)
>
> That which is proper to each thing is by nature best and most pleasant for each thing; for man, therefore, the life according to reason is best and pleasantest, since reason more than anything else *is* man. This life therefore is also the happiest. (*NE* 1178a)

Aristotle's claim that rational activity is the best and most pleasurable needs to be considered in light of what we said about the biological development of man. As we saw earlier, the capacity for reason is the "target" of human development, and hence is the most final cause for the sake of which all prior developmental stages occur. Thus the nutritive faculty develops for the sake of the desiderative faculty (*GA* 742b), and the desiderative faculty develops for the sake of the rational (*Pol.* 1334b).

Aristotle's claim that the best pleasure is associated with rational activity follows naturally from his view about human development. The best and most pleasurable activity is the activity that corresponds to the *target* of human development. The target, as we have seen, is fixed by nature and is not "up to us." Again, this is *nous* and rational activity (*GA* 736b). One looks to *that* to determine which pleasures one ought and ought not pursue. Since the target of human development is *nous* and rational activity, it follows that the pleasures of rational activity are the proper pleasures for man. So, affect is *correct* if it conduces to excellent rational activity, and it is *incorrect* if it hinders excellent rational activity. One who is driven by the apparent calling gets it wrong because the "affective content" of this calling drives one to pursue activities that do not conduce to contemplative excellence. In Aristotle's view, if one's preferred activities are not *rational* activities, then one literally lives the life of some other creature (*NE* 1178a).

The incorrectness of man's default affective state

Aristotle certainly believes that activity in accordance with affective virtue *can* and *should* be pleasurable (*NE* 1104b), but the untrained, "default setting" of the desiderative faculty opposes the correct rule (*NE* 1102a). That this is the case follows naturally from the fact that, for the first few years of life, a child's actions are caused by impulse. In numerous passages, Aristotle claims that the bodily pleasures, rather than practical wisdom, direct the actions of children (*NE* 1154b, *NE* 1179b, *NE* 1111a, *NE* 1111b). Because a child repeatedly acts at the beck and call of his bodily pleasures, he becomes accustomed to acting in this way. Since repeated actions leave their mark on one's character (*NE* 1114a), children become accustomed to acting without practical wisdom. So naturally, the "default setting" of the desiderative faculty is one that is driven by the bodily pleasures rather than by the correct rules apprehended by practical reason. Because our very first actions are driven by the bodily pleasures, we initially find it painful to submit to, and be driven by, correct rules apprehended by practical reason. Thus, the default affective state of man is one in which compliance with practical wisdom is *painful*. Obviously, this is not a very auspicious beginning for a creature for whom excellent rational activity is the good. Man begins by finding painful the very activity that he should find pleasurable.

Habituation: correcting man's default affective state

Although the default affective state of man is "anti-reason," affect can be turned round so that the individual experiences compliance with practical wisdom as something pleasurable. As we will see, cultivating affect in this way has a profound effect on the motivational system. Essentially, a cultivated affective state ensures that the motivational system is commanded by proper pleasure. I argue that without this sort of motivational system, excellent contemplative activity is impossible. As we shall see, correct affect not only provides specific motivational preconditions that are necessary for contemplative excellence, but it also provides its possessor with specific epistemological safeguards that are necessary for contemplative excellence.

Man's default affective state—indeed, his default motivational system—is driven not by man's proper pleasures, but by the bodily pleasures that are common to animals. Instead of being driven by the proper, rational pleasures, the bodily pleasures command the motivational systems of children (*NE* 1154b, *NE* 1179b, *NE* 1111a, *NE* 1111b). Consequently, children have not developed their capacity to act in a way that is unique to man. Their desiderative faculty is not directed by practical wisdom but instead is driven by the bodily pleasures. If children fail to cultivate affect and hence remain with a desiderative faculty that is not directed by the correct rule, they will live as lower animals and their actions will fail to conduce to flourishing (*NE* 1095b).

How does one transform the motivational system so that it is commanded by the correct affect? Transforming affect is not merely a "cerebral" or "intellectual" lesson that one learns in a traditional classroom setting (*NE* 1103a). This is because affect is inextricably linked to action and movement (*DA* 433a). Affect, as we noted, commands the motivational system and controls action. To "recalibrate" this system obviously requires much more than listening to a few lectures on the good. Indeed, it is the inextricable link between affect and action which leads Aristotle to criticize those who believe that merely listening to lectures and "taking refuge in theory" makes one good (*NE* 1105b). In order to cultivate affect and reverse its natural opposition to practical wisdom, a special kind of training is needed.

The public component in affective training

The training that is involved in reversing man's natural aversion to compliance with practical wisdom necessarily involves the *body* and requires the repeated performance of the actions that express affective virtue. Aristotle states that in order to acquire the affective virtues one must begin to act *as if* one already has them. Thus, the first step in cultivating affect involves repeatedly performing the actions of persons who possess a cultivated affective state. This repeated performance is a *public* performance that involves moving the body in ways recognized by the community as virtuous. The capacity to do this depends upon our capacity to recognize and imitate persons that exhibit the virtues.

Let's consider courage. Understood in purely physiological terms, a courageous act is a specific type of muscle movement that is a response to an environmental stimulus. For example, one who perceives the enemy drawing near responds courageously by engaging in a sequence of muscle movements that comprise "defending oneself." Upon sight of the enemy, the individual fetches his sword and shield and stands his ground. Aristotle says that in order to cultivate affect, we begin here. We begin by acting as the courageous man would, even though we are scared. As the character Archie Gates puts it in the 1999 film *Three Kings*, "*you do the thing you're scared of, and you get the courage after you do it, not before you do it.*" This is partly right. You get the courage after *repeatedly* doing the thing you are scared of (*NE* 1103b). One must face the enemy not once, not twice, but repeatedly, in order to develop courage. Just as repeatedly practicing a song enables you to learn it "by heart," the repeated practice of courageous actions enables you to act courageously "by heart." Repetition and practice cause such actions to become automatic and habitual and this element of stability is a necessary ingredient of a virtuous action (*NE* 1103b). In order for an action to be virtuous it must be performed from a firm and unchangeable character, and it is *repetition* of the action which leads to this stability.

In his excellent monograph addressed to teachers, *Talks to Teachers on Psychology,* William James discusses the reason why repetition leads to the automatic, habitual performance of actions. The automaticity of behavior is made possible because of the malleability and plasticity of the human nervous system:

> I believe that we are subject to the law of habit in consequence of the fact that we have bodies. The plasticity of the living matter of

> our nervous system, in short, is the reason why we do a thing with difficulty the first time, but soon do it more and more easily, and finally, with sufficient practice, do it semi-mechanically, or with hardly any consciousness at all. Our nervous systems have . . . *grown* to the way in which they have been exercised, just as a sheet of paper or a coat, once creased or folded, tends to fall forever afterward into the same identical folds.[3]

James' remarks help us to understand why the repeated performance of courageous acts results in automatic or habitual courageous action. Since the material that comprises the nervous system can be molded, the nervous system can be shaped by the behaviors in which we repeatedly engage, so as to leave a "long term memory" on our muscles. Just as repeatedly studying a concept causes the formation of a long-term memory in one's intellect, repeatedly performing an action causes the formation of a long-term memory in one's body. Aristotle thinks this fact is so obvious that he claims that only a "thoroughly senseless person" fails to realize that one's repeated actions leave their mark on one's character (*NE* 1114a).

The requirement of pleasure

Crucially, however, courageous behavior is much more than automated muscle movement. In order to be truly courageous one must *feel* a certain way, and have very specific affect. Aristotle explicitly states that one who looks courageous "on the outside" but who is petrified and pained "on the inside," is *not* truly courageous (*NE* 1104b). In addition to habitual muscle movements, one who is truly courageous *must* experience *pleasurable affect* when performing the courageous act:

> [T]he man who abstains from bodily pleasures and delights in this very fact is temperate, while the man who is annoyed at it is self-indulgent, and he who stands his ground against things that are terrible and delights in this or at least is not pained is brave, while the man who is pained is a coward. For [affective virtue] is concerned with pleasures and pains; it is on account of the pleasure that we

> do bad things, and on account of the pain that we abstain from noble ones. (*NE* 1104b)

We have said that it is only by repeatedly acting in the face of danger whereby one becomes courageous. But Aristotle seems to think that if one publicly performs in this way a certain number of times, one will take pleasure in acting courageously. But why? James' comment above explains why bodily repetition leads to muscle memory, but the question here is *why* should the repeated performance of an act cause one to eventually experience the performance of that act as something pleasurable? Why does Aristotle think that if I repeatedly stand my ground in the face of the enemy, I will eventually take pleasure in this act?

One possible answer is that habitual actions provide a source of pleasure (*Rhet.* 1370a). The idea here is that once an action becomes habitual, then it is by that very fact pleasurable. However, this answer will not do. After all, one engages in habitual action *prior to* the acquisition of courage, albeit the wrong kind. Prior to developing courageous habits, one is habitually *non*-courageous. And given that habitual actions provide a source of pleasure, it follows that the old non-courageous habits provide a source of pleasure. An explanation is needed as to *why* the newly acquired courageous habits *work* in replacing the old. Aristotle claims that given two competing pleasurable activities, one pursues the activity that yields the greater pleasure (*NE* 1175b). If courageous habits "stick," then the individual finds these habits more pleasurable than his old non-courageous habits. The question is why.

The private component in affective training

In order to answer this question, we first need to point out that repeated bodily action is not the only element involved in the development of affective virtue. Indeed, in addition to repeated muscle movements, there is another type of repetition that is equally, if not more, important. This repetition does not involve muscle movements, but instead involves something private, internal, and rational.

To see this, note that the development of affective virtue begins with a phenomenological state that is similar to that of the continent man. The continent man, says Aristotle, knows that his appetites are bad, but refuses on account of his rational principle to follow them (*NE* 1145b). The continent man is pulled in different directions but *chooses* to listen to the correct rule or "strategic plan." And this is the state from which the development of affective virtue inevitably begins. The starting point of moral development is internal tension. Discomfort is inevitable. For we are moving against our natural impulses and default state:

> [B]y abstaining from pleasures we become temperate, and it is when we have become [temperate] that we are most able to abstain from them; and similarly too in the case of courage. (*NE* 1104a)
>
> [W]e must consider the things towards which we ourselves are easily carried away; for some of us tend to one thing, some to another; and this will be recognizable from the pleasure and pain we feel. We must drag ourselves away to the contrary extreme; for we shall get to the intermediate state by drawing well away from error, as people do in straightening sticks that are bent. (*NE* 1109b)

To develop affective virtue requires that we repeatedly choose to move away from our default state and doing so is a *private* act. It is an act of will in which one chooses to listen to the correct rule or strategic plan, rather than follow the lead of mere impulse. This is inevitably hard and uncomfortable. Despite this discomfort, one must nevertheless "press on," again and again, as the rule directs. One must swim against one's own tide. This continual effort is a repeated act of will, not visible to the outside observer. It is private. This is precisely why Aristotle emphasizes that one is *responsible* for one's affective virtue (*NE* 1114a).

Affective virtue begins with the private choice to listen to a rule or rational principle *instead of* desire. The individual who chooses to develop courage says to himself, "I desire to flee, yet the rule says that I must stand my ground. I will listen to the rule." Not just once.

Not just twice. But repeatedly. The individual who chooses to develop temperance says to himself, "I am tempted and my desire is luring me to the forbidden act, yet the rule says that I must say no. I will listen to the rule." Not just once. Not just twice. But repeatedly.

It is also important to point out that the correct rule that one repeatedly listens to concerns the "intermediate state" (*NE* 1109). By this, Aristotle means a "middle ground" regarding affect, and hence "feeling" that is neither too much nor too little. Whatever this middle ground is, Aristotle observes that it will vary from person to person (*NE* 1106a–b). One person's tolerance for pain is different from another, and so the "middle affective ground" of each will be different. Nevertheless, one must be cognizant and mindful of what one's middle ground is. Thus the private component of affective training not only involves a repeated act of following the correct rule, but it also involves a *repeated mindfulness* of one's particular affective middle ground.

Now that we understand the "private" component of cultivating affect, we are in a position to understand why, once cultivated, a human being should prefer virtuous over non-virtuous habits. It is private, repeated obedience to a correct rule coupled with private, repeated mindfulness of one's affective middle ground, through which affective virtue is developed. As we shall see, Aristotle believes that through these private acts, a human being awakens a tranquility that is even more pleasurable than man's default affective state.

The awakening of proper pleasure

Recall that affective virtue and *phronêsis* are inter-entailing (*NE* 1144b). Thus, if one has affective virtue then one has practical wisdom and vice versa. *Phronêsis* is a complex ability that we will discuss in more detail a little later, but for now it suffices to note that *phronêsis* is a virtue of the rational part of the soul, specifically the part concerned with goal-oriented action (*NE* 1140b). Thus, *phronêsis* necessarily involves thinking, that is, thinking that informs and directs *action.* Indeed, the *phronimos* habitually and routinely engages in this sort of rational activity.

Because the *phronimos* has affective virtue, he has cultivated his desiderative faculty so that it habitually complies with the correct "strategic plan," directing the body accordingly. And Aristotle notes that the when the desiderative faculty is cultivated in this way, the resulting affective state is one of tranquility. As Aristotle says, the *phronimos* is no longer commanded by the bodily pleasures. Instead, the *phronimos* is commanded by a higher pleasure that does not imply pain:

> [T]he brutes and children pursue pleasures of the [body] (and the man of practical wisdom pursues tranquil freedom from that kind), viz. those which imply appetite and pain, i.e. the bodily pleasures (for it is these that are of this nature) and the excesses of them, in respect of which the self-indulgent man is self-indulgent. This is why the temperate man avoids these pleasures; for even he has pleasures of his own. (*NE* 1153a)

In fact, the actions of the *phronimos* which, again, are ones in which the correct strategic plan directs the movements of the body, yields a pleasure that is proper and unique to human beings:

> Neither practical wisdom nor any state of being is impeded by the pleasure arising from it; it is foreign pleasures that impede, for the pleasures arising from thinking and learning will make us think and learn all the more. (*NE* 1153a)

Because practical wisdom is a virtue of the rational part of the soul, the pleasure that arises from it is a proper pleasure. For the rational pleasures are the proper pleasures of man (*NE* 1178a). Indeed, it is because practical wisdom yields a proper pleasure that explains why Aristotle says that virtuous actions are pleasant by nature (*NE* 1099a). If properly habituated, a human being *must* experience the activity of practical wisdom as pleasurable. Why? Because practical-wisdom produces rational pleasure and human beings possess an innate preference system such that they *must* delight in the rational pleasures. Just as a human being must experience pleasure from eating, drinking, and sexual activity, so too a human being must experience pleasure from rational activity. The rational pleasures are no less human than the bodily pleasures (*NE* 1111b).

But not only *must* a human being delight in the rational pleasures, but if properly cultivated, a human being will *naturally prefer* the rational over the bodily pleasures. For one prefers the pleasure of the better part (*Pol.* 1333a). Since the body exists for the sake of rational activity (*Pol.* 1334b), the rational pleasures are the pleasures of the better part. Given that a human being naturally prefers the rational pleasures, then, if properly cultivated, rational pleasures, rather than the pleasures of the body, will *command* the motivational system. And an individual who is commanded by the rational pleasures is one who pursues rational activities. To be commanded by the rational pleasures is to motivated—indeed, *intrinsically* motivated—to pursue the activities that yield these pleasures. This describes the *phronimos*: he has tasted a higher, rational pleasure, and it is this pleasure that commands him.

Yet, most think that the bodily pleasures are the *only* pleasures (*NE* 1153b). In Aristotle's view, most have motivational systems that are commanded not by the proper pleasures of man, but instead by pleasures common to cattle (*NE* 1095b). But why is this, if, after all, human nature is built to delight in rational pleasure? The answer is that man's proper pleasure must be *awakened*, *stimulated*, and *nurtured* so that it, rather than some other pleasure, commands the motivational system.

Because the typical individual does not undertake the effort and determination requisite to the cultivation of affect, he fails to fully awaken his proper pleasure. As Aristotle says, most have "never tasted the pure and generous pleasure" that comes from rational activity (*NE* 1176b). The many, he says, live by the bodily passions because "they have not even a conception of what is noble and truly pleasant, since they have never tasted it" (*NE* 1179b). The implication here is that the pleasure arising from affective virtue/*phronêsis* must be awakened. Although all have the potential to awaken this proper pleasure, to do so requires deliberate work, energy, and effort. Again, this is the hard work of repeatedly (and mindfully) swimming against one's tide and listening to the correct rule. It is as if proper pleasure is a frozen sea within us, and each mindful act of listening to the correct rule is an axe blow to the sea.[4] Through repeated private acts of mindful compliance with the correct rule, we awaken a proper pleasure—the pleasure of having our actions directed by thought.

The role of law

Most people do not and will not suffer the hard work required to awaken proper pleasure. This is not surprising since the work is painful and difficult, and the natural instinct is to avoid pain (*NE* 1172a). Combined with the magnetic pull of our default affective state, the likelihood that someone will endure the effort required to transform affect is slim indeed. Nor does a parent have the requisite authority to command his child to do this sort of work (*NE* 1180a). And this is precisely why Aristotle places the cultivation of affect within the province of law and legislation. For law has the power to compel, and even though compulsion is painful (*Rhet.* 1370a), when the law is correct we do not mind being compelled by it (*NE* 1180a). Sadly, however, Aristotle complains that most states do not use the law to cultivate affect:

> In the Spartan state alone, or almost alone, the legislator seems to have paid attention to questions of nurture and occupation; in most states such matters have been neglected, and each man lives as he pleases, Cyclops-fashion, "to his own wife and children dealing law." Now it is best that there should be a public and proper care for such matters. (*NE* 1180a)

Despite its nearly universal inefficacy, Aristotle nevertheless claims that the state should be involved in cultivating affect because, when left to their own devices, human beings naturally avoid the hard work necessary to cultivate affect. Without the nourishment provided by law, it is unlikely that a human being will awaken his proper pleasure. And if the human being fails to awaken this pleasure, he is saddled with a motivational system commanded by the wrong affect and therefore driven by an apparent or false calling. The proper role of the state is to prevent this state of affairs by nurturing the motivational system so that it is driven by the correct afffective content, thereby allowing its possessor to flourish.

The developmental role of proper pleasure

Now, the proper pleasure that affective virtue introduces into the motivational system is not the *only* proper pleasure. For practical wisdom is not the only kind of rational activity, and each type of rational activity has its own corresponding pleasure (*NE* 1175a–b). The activity of practical wisdom is altogether different from contemplative activity. Practical wisdom is concerned with the variable and the contingent that characterize human affairs, whereas contemplative activity is concerned with the invariable and the necessary:

> [T]here are two parts which grasp a rational principle—one by which we contemplate the kind of things whose originative causes are invariable, and one by which we contemplate variable things. (*NE* 1139a)

Aristotle states that the pleasure arising from contemplative activity is the highest and best pleasure (*NE* 1178a). And it is this claim that is at the crux of the intellectualism debate. Some say that the pleasures of contemplation can be severed from, and be pursued independently of, the pleasure of ethical virtue. But as we shall see, ethical virtue is *required* for excellent contemplative activity and cannot be severed from it. One can only become *sophos* if one has a certain motivational system, and ethical virtue *is* that system.

The Development of Contemplative Excellence

Excellent contemplative activity is a masterful rational activity that builds upon the inner resources implied by practical wisdom. Like affective virtue, excellent contemplative activity takes training and time. In the *Metaphysics*, Aristotle discusses some of the generally accepted views about the *sophos*. The wise person knows all things in a general way, knows the most controlling science and therefore has an understanding of the final causes of things (*Met.* 982a–b). Coming to know such things is a lengthy process that begins with perception, depends on true belief, and requires a dialectical consideration of competing theories. We will discuss this process in more detail in a moment, but the important point for now is that contemplative excellence does not happen overnight, but arises only if one undertakes the necessary intellectual work. In fact, there are two preconditions to developing the abilities on which contemplative excellence depends: one affective, the other cognitive. Ethical virtue implies both.

The motivational preconditions of contemplative excellence

Aristotle explicitly acknowledges that the pleasure that one finds in an activity motivates one to pursue that activity. In fact, he acknowledges that those who excel at an activity do so *because* they find the activity pleasurable:

> [A]n activity is intensified by its proper pleasure, since each class of things is better judged of and brought to precision by those who engage in the activity with pleasure; e.g. it is those who enjoy geometrical thinking that become geometers and grasp the various propositions better, and, similarly, those who are fond of music or of building, and so on, make progress in their proper function by enjoying it; so the pleasures intensify the activities, and what intensifies a thing is proper to it. (*NE* 1175a)

When one finds an activity pleasurable, one is motivated to pursue that activity. And when one is motivated to pursue an activity one excels at that activity. Aristotle further acknowledges that the converse is true as well. If one finds an activity to be painful, one avoids the activity. And of course, if one avoids an activity one will not excel at that activity:

> [A]ctivities are destroyed by their proper pains; e.g. if a man finds writing or doing sums unpleasant and painful, he does not write, or does not do sums, because the activity is painful. (*NE* 1175b)

Clearly, if one is to become excellent at contemplation, one must find rational activity pleasurable. If one does not take pleasure in thinking and learning, one will not pursue the actions that lead to contemplative excellence. On the other hand, if one finds thinking and learning pleasurable, one will be compelled to pursue these activities. As Aristotle says, the pleasures arising from thinking and learning will make us think and learn all the more (*NE* 1104b). And this is precisely what ethical virtue guarantees. For ethical virtue implies that its possessor is commanded by rational pleasure and hence is motivated to pursue rational activities. This is precisely the affective state and the motivational orientation that leads to contemplative excellence. Intrinsically motivated to pursue rational activities, the ethically virtuous individual is an ideal candidate to become *sophos*.

The cognitive preconditions of contemplative excellence

Contemplative excellence not only requires that one is desirous of rational activity, but it also has intellectual or cognitive preconditions as well. *Sophia*, which is the virtue responsible for excellent contemplative activity, is defined as a combination of *nous* and *epistêmê* or knowledge (*NE* 1141a). "Knowledge" for Aristotle is not merely the possession of "justified, true belief." *Epistêmê* or knowledge is something *very* specific for Aristotle. Not just any statement can be known. In Aristotle's view, in order to know some claim P, (1) P must be necessarily true, and (2) one must be able to *demonstrate* or prove P from premises that are necessarily true and universal (*Post. An.* 71b, 88b). This implies that P is universal statement. A "universal statement" is simply a statement about a class or category of things, rather than a statement about a particular entity. Thus "all men are bipeds" and "all prime numbers are odd" would qualify as universal statements, whereas the statement "Socrates is a man" would not.

Aristotle's definition of knowledge implies that claims like "Socrates is a man," and "the house is on fire," are not objects of knowledge. Why? Because only universal claims can be "known," and the above two claims are particular, rather than universal claims. Aristotle would say that one *perceives* that Socrates is a man, and one *perceives* that the house is on fire. But Aristotle distinguishes perception from knowledge—animals and children have perception, but they do not have knowledge (*DA* 415a). Aristotle acknowledges that perception can be correct or incorrect (*DA* 428b), but he refrains from calling correct perception *knowledge*. Again, if knowledge of P requires that P must be derived from necessarily true universal premises (*Post. An.* 71b, 88b), then P can never be a particular statement like "Socrates is a man."

The origin of universals

In Aristotle's view, perception is of the particular, whereas knowledge is of the universal. However, knowledge begins with, and develops out of, perception. Aristotle is an empiricist, and views all knowledge as ultimately arising from sense perception combined with memory:

> So from perception there comes memory, as we call it, and from memory (when it occurs often in connection with the same thing), experience; for memories that are many in number form a single experience. And from experience, or from the whole universal (the one apart from the many, whatever is one and same in all those things), there comes a principle of skill and of understanding—of skill it if it deals with how things come about, of understanding if it deals with what is the case. (*Post. An.* 100a)

For Aristotle, the universals on which contemplative activity operates, originate from a raw material that is an effect of perception, namely, memory. Memory, according to Aristotle, is a "residue" caused by perception. The operation of the world on our sense organs leaves a sort of imprint or residue on our consciousness, i.e., memory. If this imprint is repeated a sufficient number of times "experience" emerges. Shortly after the passage quoted above, Aristotle describes experience as a "primitive universal." Thus, experience is the "basic material," the "primitive universal" from which all scientific knowledge begins. As Aristotle says, science and craft arise *out of* experience (*Met.* 980b). Just as Michelangelo's *David* began as a slab of un-worked marble, knowledge and understanding begin with similar "raw materials," namely the primitive universals that comprise experience. One acquires understanding and wisdom by working on these materials with the intellect.

The neurobiology of memory

As we begin to understand Aristotle's account of knowledge, it is interesting to note that his account has deep similarities to the current understanding about the relationship between memory and learning. Aristotle regards perception and memory as bodily processes. In the *De Anima,* he claims that the faculty of sensation is dependent upon the body (*DA* 429b), and in *On Memory,* he claims that memory is "of the body" (*somatikon*). Recollecting (the attempt to remember) consists in searching for an 'image" in a corporeal substrate (*Mem.* 453a). Thus, knowledge begins with the body and bodily processes.

Research on the "neurobiology of memory" (the study of the cells involved in memory) demonstrates that any form of learning—whether short or long term—involves the physical cells that comprise the nervous system.[5] Memories have a physical, cellular basis in the brain and therefore depend on the body. Aristotle's belief that knowledge begins with the bodily processes involved in memory is alive and well in contemporary neuroscience.

Scientists studying the neurobiology of memory have discovered that each and every experience causes a biochemical event in the brain. A "biochemical event" is simply a complex reaction encompassing a cascade or sequence of smaller events involving the brain's chemistry. So when I show you a picture of an elderly bald headed man with a snub nose and say "Socrates," a corresponding biochemical event occurs in which "chemical messengers" (neurotransmitters) send information to cells. When the cells "receive" this information, they release further chemicals, and so on. Think: chain reaction. True, when you see the picture of the old man and hear "Socrates," you experience this as a single (rather uninteresting) event. However, what is *actually* happening in your brain is a complex chain reaction involving a host of biochemicals. This complex biochemical event is tantamount to the formation of a neural network, that is, a relationship among neurons. Neurons are the cells that comprise your brain. They not only store information but they also communicate with each other through *synapses.*

Aristotle claimed that sensation leaves a "residue" which he referred to as memory, and that the same memory repeated a sufficient number of times—thereby building up this "residue"—yields "experience" or the "primitive universal." The essentials of that story are the same within the modern account of memory, except the residue left by sensation is known by contemporary scientists as a *neural network.* So each time you see the picture of the old man and hear "Socrates," the "residue" is a neural network. The question is, what happens when this same neural network is repeated a sufficient number of times? What happens when we repeatedly hear "Socrates" whenever we see a certain picture of an elderly bald headed man? What happens when the built-up residue reaches a "critical mass"? Aristotle claimed that such repetition gave rise to something new, namely, "experience." The essentials of this story remain intact in the modern scientific account.

When a neural network is repeated a certain number of times—when the "residue" reaches a critical mass—a gene for growing bigger, stronger synapses is activated or "turned on."[6] Through practice and repetition one awakens genes—switches, really—that are responsible for building bigger, stronger connections among the cells in the brain in which memories are stored. And these bigger, stronger connections are associated with long-term memory and hence deep, lasting learning. Thus, the "new" thing that repetition of a neural network creates is the awakening of a gene that builds stronger connections and hence transforms the neural network into a long-term memory. Through repetition and practice one literally changes one's brain at the cellular level.

As you may have already surmised, the neuroscientific account of memory also helps us to understand what is happening at the neurobiological level when one repeatedly engages in the acts required for the development of affective virtue. Both the repeated public performance of courageous acts, and the repeated private act of listening to the correct rule, awaken a gene that builds a stronger neural network, and hence a long-term memory. After repeatedly performing a certain number of courageous actions, such actions occur automatically because the neural network responsible for "courageous action" is strong and solid.

With this primer in the neurobiology of memory out of the way, let's return to our discussion of Aristotle's account of knowledge. Thus far we have seen that knowledge—which implies the ability to prove true, universal claims—originates with the bodily processes of perception and memory. These natural physiological processes give rise to the primitive universals from which all other universals originate (*Met.* 981b). The formation of primitive universals happens quite naturally and no argument or rational process is responsible for their existence. Just as your body's digestive system operates independently of your conscious thoughts, so too does the formation of primitive universals. They just happen.

The activities on which knowledge depends

However, the acquisition of knowledge does *not* just happen. The acquisition of knowledge requires one to consciously and deliberately engage in intellectual activity. Experience and "primitive universals" happen naturally, but experience is not the same as knowledge (*Met.* 981a). Knowledge begins and grows out of experience (*Met.* 981b). Only by engaging in some sort of intellectual activity can one acquire knowledge. Now, there is some scholarly debate about the precise nature of the intellectual activity that is required for knowledge. For Aristotle discusses two types of activity, *induction* (*Post. An.* 100b) and *dialectic* (*Top.* 100a–b) which are involved in the acquisition of the "first principles" or starting points of knowledge. My own view, a complete defense of which is beyond the scope of this book, is that both induction and dialectic are involved in the acquisition of knowledge, albeit at different moments.

To begin to understand the type of intellectual work necessary for knowledge, we need to be mindful of the fact that knowledge is limited to "universal propositions." And importantly, the *subject* of a universal proposition will always be a universal. What do I mean? Consider the following two sentences:

A. Socrates is a man
B. All men are mortal

In sentence A, the subject is "Socrates" who is a particular, tangible, specific individual. In sentence B, however the subject is "All men" which is not a particular, tangible, specific, individual. In sentence B, the subject is a universal—the class of all men. Only sentences like sentence B can be demonstrated, and therefore only sentences like B can be *known*. Sentences like A are perceived but are not known. To help you appreciate the difference between sentences A and B, column A lists examples of statements about particulars, and column B provides examples of statements about universals:

A	B
1. Finster is my dog	1. All dogs have ears
2. This book is old	2. Books have pages
3. Socrates was Just	3. Just people are virtuous

Note that sentences like those in column B express information that is different in kind from the information expressed in column A. Sentences such as those in column B express a relationship between a universal and a universal. However, sentences such as those in column A express information about the relationship between a *particular* and a universal. Again, only sentences like those in column B are potentially knowable. Sentences like those in column A are perceived or opined, but cannot, strictly speaking, be known.

In order to acquire true beliefs that have the form of the statements in column B, one must do some intellectual work. Specifically, one must engage in either induction or dialectic. About induction, Aristotle writes:

> [I]nduction is a passage from individuals to universals, e.g., the argument that supposing the skilled pilot is the most effective, and likewise the skilled charioteer, then in general the skilled man is best at his particular task. Induction is the more convincing and clear: it is more readily learnt by the use of the senses. (*Top.* 105a)

Induction is an intellectual process that presupposes the observation of patterns and relationships among primitive universals. Note that the particular judgment "the skilled pilot is the most effective" presupposes the primitive universal "skilled pilot." By observing a similarity between two primitive universals, "the skilled pilot" and "the skilled charioteer," (i.e., that both are effective in their respective trades) one is led to the more general claim that "the skilled man is effective." In the above passage, Aristotle seems to be describing some sort of analogical reasoning. And through this sort of rational activity, one is led from primitive universals to higher order universals. One can then use these new found universal judgments as stepping stones to judgments of an even higher order, and in this way proceed to judgments about the highest universal of all, which Aristotle identifies as being *qua* being, to which the science of metaphysics or "first philosophy" is devoted (*Met.* 1026a). But again, even the highest universal is apprehended only after one undertakes the necessary intellectual work that enables one to move from a grasp of primitive universals to a grasp of universals of a more general, higher order. Induction is a way by which one undertakes this intellectual work.

Dialectic is another intellectual process by which one can arrive at the starting points for knowledge. Essentially, dialectic is a process of sorting and clarifying the relevant opinions or *endoxa* on the subject. According to Aristotle, *endoxa* are opinions accepted by the majority, by wise persons, or by craftsmen (*Top.* 100b). Dialectic proceeds by determining the strengths and weaknesses of opinions by comparing and contrasting them with other opinions, by detecting ambiguities, and by drawing relevant and important distinctions.[7] From this process, first principles can emerge. There are many instances in which Aristotle engages in dialectic. In the *De Anima,* he considers various theories of the soul and subjects them to scrutiny; in the *Nicomachean Ethics,* he scrutinizes various theories of the human good; and in the *Physics,* he scrutinizes various theories of change and reality. Through such criticism, Aristotle distills a stronger, more coherent position that provides a starting point for further inquiry.

Now that we understand the different intellectual processes involved in the acquisition of knowledge, let's quickly review the various elements on which the development of knowledge depends. Knowledge depends upon the natural processes associated with perception and memory. These natural processes provide the "raw material," the primitive universals, on which all understanding depends. With this raw material in place, one can then engage in induction or dialectic. One can begin by seeing patterns among primitive universals that provide evidence for more general claims about these universals. One then sees patterns among these more general claims, and so on. Then, one further refines one's beliefs by considering the *endoxa* and arriving at more universal claims. Ultimately, scientific knowing requires that one can prove that these universal claims follow logically from other general claims that are necessarily true.

Securing the Development of Contemplative Excellence

Crucially, the path from perception to knowledge can go wrong in many ways. Aristotle himself notes that most people are unable to make important distinctions (*NE* 1172b). In addition, people can misjudge the strengths or weaknesses of opinion, as well as misjudge the veracity of an opinion (*NE* 1172a–b). Clearly, misjudging the strength of an opinion would hamper one's ability to engage in dialectic. One needs to know *what* to criticize and what to praise—what portion of an opinion is true, and what portion of an opinion is false—in order to use dialectic to arrive at first principles. Making matters worse is the fact that the majority of people have false opinions (*Rhet.* 1378a). This sort of affliction would doubtless hamper induction. Imagine, for instance, if one had false beliefs about the skilled charioteer and the skilled pilot. These false beliefs would further infect the more general beliefs about these universals that are arrived at via induction, and so on. Even at "ground zero" of knowledge, that is, *perception,* there can be error. For the perception that the white thing is this or that may be false (*DA* 428b). In other words, I can mistakenly characterize the white thing as *limestone,* when in fact it is *quartz.* As Aristotle says, there are many ways to go wrong (*NE* 1106b), and as you can see, this holds true for the acquisition of knowledge.

We have noted that above that ethical virtue implies the motivational orientation towards rational activity that is necessary for contemplative excellence. One who possesses ethical virtue has awakened a proper, rational pleasure, takes pleasure in rational activity, and therefore will be desirous of more rational activity (*NE* 1104b). But being motivated to pursue rational activity is not sufficient for *excellent* rational activity. Given the numerous errors that can trip up one's search for knowledge, one needs to avoid these errors in order to realize contemplative excellence. As we shall see, because ethical virtue implies *phronêsis,* one who possesses ethical virtue is protected from these errors.

Phronêsis

Phronêsis is an excellence of practical reason and implies a number of epistemic strengths. One of these strengths is excellence in practical deliberation (*NE* 1142b). As mentioned earlier, this ability is an excellence in strategic planning and the type of practical thinking involved in goal-directed behavior.

Because he possesses affective virtue, the *phronimos* is driven by desire for rational activity, and he therefore uses strategic planning to direct the body in order to realize this desire. But the sort of practical thinking in which the *phronimos* engages is not just some sort of algorithmic decision procedure. Aristotle also says that *phronêsis* deals with the contingent, the uncertain, and the variable (*NE* 1143a). The context of practical affairs is dynamic, unpredictable, and cannot be mastered within the confines of universal, absolute rules (*NE* 1137b). Circumstances and context *matter* and determine the specific means that one should adopt to realize rational activity. As Aristotle says, sometimes it is *unhealthy* to engage in contemplation (*NE* 1153a). Thus, the *phronimos* uses practical deliberation to decide just what, given his present circumstances, is the best way to engage in rational activity. It could be "contemplate, right here, right now" or it could be "stand your ground and face your enemy." But whatever the circumstances, the *phronimos* makes the right decision.

Success in navigating the landscape of practical affairs requires a number of special epistemic skills different from those used within, for example, geometry or physics. Absolute rules are instruments much too blunt to capture the relevant circumstantial factors. *Phronêsis,* therefore, implies masterful discernment. As Aristotle says, it is an excellence of perception (*NE* 1142a). He also says that *phronêsis* is the excellence of the opinion making faculty (*NE* 1140b). Thus, the *phronimos* has true beliefs and this explains why he his excellent at deliberating. But again, in the domain of practical affairs, excellent deliberation and opinion are sensitive to the circumstances and context of the situation. Accordingly, Aristotle also mentions several component abilities that are implied by the excellence of opinion in the practical realm. Among these abilities are *sunêsis* and equity (*epikeia*). *Sunêsis* is a virtue which enables one to be a good judge of another's opinion (*NE* 1143a), whereas equity enables one to correctly perceive the factual circumstances in which a rule does not apply (*NE* 1137b). Clearly, *phronêsis* is a master intellectual ability. It implies the ability to realize rational activity in a messy domain of shifting circumstances embedded in unpredictability.

But the various epistemic strengths possessed by the *phronimos* are precisely those strengths that protect against the numerous obstacles in the acquisition of scientific knowledge. Consider, for instance, *sunêsis*. *Sunêsis* implies that its possessor knows how to correctly sharpen an opinion and detect the strengths and weaknesses of another's opinion. Equity, moreover, implies an ability to correctly identify the factual circumstances which demand the modification of a rule. These strengths are precisely those abilities that enable one to engage in dialectic with excellence.

Moreover, the fact that Aristotle identifies *phronêsis* as the excellence of the opinion making faculty (*NE* 1140b) is a clear indication that this ability is essential to the acquisition of knowledge. Plainly, knowledge must begin with true opinion and this is precisely what *phronêsis* guarantees. Art and science grow out of experience, and one must begin with the "fact that" before one knows the "reason why" (*Met.* 981a–b). Insofar as it implies the correctness of the "fact that," *phronêsis* provides the starting point for knowledge.

In fact, affective virtue, which *phronêsis* implies, also protects against some of the mishaps that can occur in the acquisition of knowledge. Affective virtue precludes a number of distorting cognitive effects that result from excessive affect. Consider what Aristotle says about how affect can distort the natural bodily processes from which knowledge originates. In the following passage, Aristotle claims that those who are overly passionate are unable to retain the impressions (memories) that are necessary for the development of experience:

> The process of movement [sensory stimulation] involved in the act of perception stamps in, as it were, a sort of impression of a percept, just as persons do who make an impression with a seal. This explains why, in those who are strongly moved owing to passion, or time of life, no mnemonic impression is formed; just as no impression would be formed if the movement of the seal were to impinge on running water; while there are others in whom, owing to the receiving surface being frayed, as happens to [the stucco on] old [chamber] walls, or owing to the hardness of the receiving surface, the requisite impression is not implanted at all. Hence both very young and very old persons are defective in memory (*Mem.* 450a).

Excessive passion adversely affects the function of memory, which, as we now know, is essential to knowledge and understanding. If no "mnemonic impression" is formed, then the universals on which understanding depends will not emerge. Excessive passion therefore limits the "raw materials" on which scientific knowledge depends.

Excessive passion not only impairs the function of memory, but it also distorts judgment. Consider some of Aristotle's remarks from the *Rhetoric:*

> It is not right to pervert the judge by moving him to anger or envy or pity—one might as well warp a carpenter's rule before using it. (*Rhet.* 1354a)

> [The members of the assembly] will often have allowed themselves to be much so influenced by feelings of friendship or hatred or self-interest that they lose any clear vision of the truth and have their judgement obscured by considerations of personal pleasure or pain. (*Rhet.* 1454b)

> The emotions are all those feelings that so change men to affect their judgments, and that are also attended by pain or pleasure. (*Rhet.* 1378a)

In these passages, Aristotle states that opinion and belief can be unduly influenced by affect. And in Book VII of the *NE*, where Aristotle discusses weakness of will or *akrasia,* he further observes that affect can distort *perception* (*NE* 1147b). Akratic individuals act under the influence of the passions, and are in a similar condition to those who are asleep, mad, or drunk. Such individuals possess the correct normative principles but their bodily passions distort their perceptual judgment. So the although the akratic believes that "sweet things should not be tasted," his passion and desire prevent him from recognizing the cookie *as* something sweet, and he proceeds to eat it. But the *phronimos* is not subject to distorting of effects of either excessive or wrong affect. After all, the *phronimos* possesses affective virtue, and has excellent perception (*NE* 1142a).

Ethical virtue: nourishing and safeguarding contemplative excellence

Clearly, ethical virtue, which contains both *phronêsis* and affective virtue, implies a number of "inner resources" that are necessary for the development of contemplative excellence. Affective virtue not only protects one from the distorting effects of excessive affect, but also provides the correct motivational orientation to rational activity. *Phronêsis,* moreover, provides its possessor with true beliefs, masterful discernment, and an excellence in thinking. The combination of affective virtue and *phronêsis* supply the necessary inner resources—the motivational and cognitive preconditions—necessary for excellent contemplative activity.

Understanding that ethical virtue implies the inner resources that are necessary for excellent contemplative activity, enables us to see *why* Aristotle believed that affective virtue and *phronêsis are* virtues. A virtue is what makes the work or *ergon* of a human being be done well (*NE* 1106a). And the *ergon* of man, as we have seen, is rational activity (*NE* 1097b–1098a). Insofar as affective virtue immunizes its possessor from the corrupting effects of excessive affect, it clearly conduces to excellent rational activity. And insofar as affective virtue implies that its possessor finds correct rational activity pleasurable, it conduces to excellent rational activity in virtue of its motivational power. Finally, insofar as *phronêsis* implies excellent opinion and perception (from which all knowledge begins), *phronêsis* clearly conduces to excellent rational activity.

The intellectualism debate revisited

In failing to understand that ethical virtue implies the inner resources on which contemplative excellence depends, intellectualists have wrongly claimed that one could be *sophos*, without possessing ethical virtue. Rather than understanding that ethical virtue is essentially *the* motivational system that *must* exist if an individual is to become a *sophos*, intellectualists have portrayed ethical virtue as an essentially "other regarding" virtue.[8] This misconception of ethical virtue has caused intellectualists to claim that ethical virtue could sometimes be incompatible with *sophia*.[9] Indeed, *if* ethical virtue is essentially concerned with our commitment to others, and *if* flourishing consists solely in theoretical activity, then it is not difficult to imagine cases in which one's concern with others conflicts with one's pursuit of *theoria*. However, this misconstrues Aristotle's account of ethical virtue. To be sure, Aristotle surely believes that family, friends, and community are valuable (*NE* 1169b–1170b, *NE* 1123a), and he even claims that a man ought to help his friends and children become good if they are living under ignoble laws (*NE* 1180a). However, for Aristotle, the essence of ethical virtue does not consist in one's commitment to other people. For Aristotle, ethical virtue is primarily "self" as opposed to "other" regarding.

Ethical virtue is essentially an excellence of a human being's *motivational system*: affective virtue is the excellence of the desiderative component, and *phronêsis* is the excellence of the intellectual component. A motivational system that possesses ethical virtue contains the inner resources that prime its possessor to engage in the actions that lead to contemplative excellence. Such a motivational system causes one to perform the specific actions through which the calling of man is realized. Hence, such a motivational system is aligned with the target of nature. To be sure, ethical virtue is not the *complete* maturation of the motivational system, but it is the motivational system that is *necessary* for this complete maturation.

Obviously, understanding ethical virtue in *this* way portrays it as something much different than an "other regarding" virtue.

The Development of the Students

In addition to misrepresenting Aristotle's notion of ethical virtue, the parties in the intellectualism debate have also paid insufficient attention to the remarks that Aristotle makes about proper student of the *NE*. Aristotle's remarks explicitly state that the *NE* is written for a very specific type of student, namely, one who has a "proper upbringing" and hence acts not from the bodily passions but from reason (*NE* 1095a). Thus, the *NE* is not written for just anyone, but only for the student who possesses the appropriate "inner resources." Aristotle states that the *NE* brings great benefit to such a student (*NE* 1095a). What is this benefit? I believe that it is the sharpening and refining of the *content of the calling* that drives the actions of the student. Like a finishing tool used by a craftsman, Aristotle intended the *NE* to further refine the affective content of the student's calling—the ultimate object of desire—that drives the student's motivational system.

The transformative aim of the *Nicomachean Ethics*

The *NE* is intended to be a transformative tool. Its purpose is not to *know* what flourishing is but to flourish—to "become good" (*NE* 1103b). The end is not knowledge but action (*NE* 1095a). Aristotle, therefore, intended the *NE* to affect the behavior of his audience. But not every individual is moved by words and argument (*NE* 1179b). Only those with a specific motivational system and specific affect are able to steer their choices and actions by argument and teaching (*NE* 1179b). This is why Aristotle lays out as a requirement that the student is not governed by the bodily passions. In order for the objective of the *NE* to be achieved, its student must be the type of individual who acts not from emotion and impulse, but from reason and argument. Thus, the student has cultivated his affective state and possesses the appropriate "inner resources" that enable his behavior to be affected by the *NE*. The student's actions proceed not from his default motivational system, but from a cultivated and refined motivational system driven by the pleasure of correct rational action.

Importantly, reason and argument are not the only means by which objective of the *NE* is achieved. For the *NE* also uses the affect of the student to achieve its aim. *The man who uses affect well will be good, and he who uses it badly will be bad* (*NE* 1105a). Indeed, Aristotle uses the affect—the pleasures—of his students to engender the maturation of their motivational systems. It is a given that Aristotle's students find correct rational activity pleasurable and have therefore awakened the proper pleasure of man. And working with this fact, Aristotle exposes the students to a form of rational activity that is different from the type of rational activity to which the student is accustomed.

The student of the *NE* has internalized the norms of the community, but it is doubtful that he has reflected on his actions. It is doubtful that the student understands *why* his actions are virtuous. Importantly, it is the student's *own* actions that provide the starting point for inquiry in the *NE* (*NE* 1095b). The *NE* exposes the student to a new kind of rational activity that has a different feel and flavor than the ethically virtuous activity with which he is familiar. This new rational activity is *self-reflective* and is concerned with "the why." Looking at his actions as *data,* the student considers his actions as a starting point for inquiry. And it is *because* the student has a specific type of motivational system that enables him to be profoundly impacted—indeed moved—by this investigative activity. For this self-reflective investigative activity is a *rational* activity. And because the student has awakened his proper pleasure, he will take pleasure in this activity. *The pleasures arising from thinking and learning will make us think and learn all the more* (*NE* 1153a).

Aristotle achieves the objective of the *NE*—to make his students good—by exposing them to a different rational activity, that is, doing "ethical and political theory" or work in the practical sciences. This rational activity has its own distinctive pleasure that is different from the pleasure associated with, e.g., courageous action. And because the student is commanded by the pleasure associated with rational action, he will delight in the different kind of rational activity in which "ethical theory" consists.

By exposing the student to the pleasures of self-reflective activity, the *NE* helps the motivational system of its pupil to mature and evolve, further cultivating the inner resources that will motivate the student to perform the actions that lead to contemplative excellence. Put most simply and crudely, at the end of the *NE,* the student realizes "wow, that was fun!" thereby becoming desirous of more opportunities to engage in this type of rational activity. By engaging and nurturing the student's desire for rational activity, and by exposing him to a new *type* of rational activity, the *NE* helps to direct the actions of the student so that he is more likely to develop *sophia.* Later on, we will see that the students' self-reflection in the *NE* also deepens their appreciation of community, and that this, too, helps his students to "become good." But the important point for now is to recognize that the *NE* was intended to engender the moral development of its students by providing them with an experience designed to arouse and delight.

Conclusion

In this chapter, I have shown that ethical virtue is a developmental prerequisite for contemplative excellence. Both intellectualists and inclusivists have misconstrued the nature of ethical virtue and have overlooked the nature of Aristotle's project in the *NE*. Ethical virtue is not merely some "other regarding" virtue but is the possession of a specific motivational system, one that is necessary for excellent contemplative activity. Indeed, ethical virtue (the combination of affective virtue and *phronêsis*), is the *starting point* for contemplative excellence. *Becoming* an excellent contemplator is possible only if specific motivational and epistemological conditions are present, and ethical virtue implies these conditions or "inner resources."

The failure to understand ethical virtue as the possession of a specific motivational system has led intellectualists to wrongly claim that one could be *sophos*, without possessing ethical virtue. For rather than understanding ethical virtue as an excellent motivational system, intellectualists speak of ethical virtue as being essentially concerned with one's commitments to other people.[10] This misconception of ethical virtue has caused intellectualists to claim that ethical virtue could sometimes be incompatible with *sophia*.[11] Indeed, *if* ethical virtue is essentially concerned with our commitment to others, and *if eudaimonia* consists solely in theoretical activity, then it is not difficult to imagine cases in which one's concern with others conflicts with one's pursuit of *theoria*. But ethical virtue, as I have shown, does not merely consist in the excellence of our behavior towards others. Rather, it is the motivational system that one needs to *become* an excellent contemplator. Thus, intellectualists are wrong in claiming that ethical virtue and *sophia* could come into conflict with one another. For *sophia* needs ethical virtue in order to be realized.

Aristotle's project in the *NE* is to help his students *become* excellent contemplators. Excellent contemplative activity is the cumulative result of specific actions and decisions that are the "outputs" of a motivational system. Aristotle's project in the *NE* is to nurture and cultivate the students' motivational system so that it is more likely to produce these outputs. He does this by exposing the student to a type of rational activity distinct from ethically virtuous activity in an effort to arouse and stimulate the students' desire for rational activity—which ethically virtuous activity implies. In the *De Anima*, Aristotle suggests that the realization of man's true calling—contemplative excellence—has a "trickle down" effect on ethical virtue, further fortifying and strengthening it (*DA* 429b). Just as ethical virtue secures and buttresses the development of contemplative excellence, contemplative excellence further secures and buttresses ethical virtue. Put another way, once realized, contemplative excellence exists in a symbiotic relationship with ethically virtuous activity—the two nurture and fortify each other. Rather than being incompatible with ethical virtue as the intellectualists contend, contemplative excellence arises out of, strengthens and fortifies ethical virtue.

By providing his students with a new form of pleasurable rational activity designed to lead them to contemplative excellence, Aristotle helps his students to become good. Ultimately, Aristotle guides the student so that he can engage in the actions through which his "true calling" is realized.

In the following chapter, I show that even in our post-modern world which rejects the idea that there is a single activity that comprises the true calling of man, the fundamentals of Aristotle's theory are both defensible and compelling.

Chapter 4

Aristotle's Developmentalist Approach Modernized: Flourishing as Self-Concordance

Is Aristotle's Theory of Flourishing Still Relevant?

Although Aristotle's ethical theory is regarded as one of the most important contributions to moral theory, Aristotle is often criticized for singling out contemplative activity as the highest and best activity in which a human being can engage. Some read Aristotle as guilty of a sort of "philosophical imperialism." According to this criticism, Aristotle assumes that a life replete with *his* favorite activity—philosophical activity—is the happiest life for human beings. But, so the criticism goes, there are thousands of activities *other than* philosophical contemplation that are beloved by their participants. Some people love the competition of sport and derive the most pleasure from physical challenges. Others love to create things with their hands. Certain others find their bliss in playing a musical instrument. And still others love to be outdoors, to garden, to hike, and so on. For many, philosophical activity is not ranked as the most pleasurable and enjoyable activity. In fact, there certainly are individuals who find philosophical contemplation rather boring. Is Aristotle's theory of human flourishing irrelevant because of its heavy emphasis on excellent rational activity and contemplation?

Granted, singling out *theoretical contemplation* as the activity that is most pleasurable to a properly functioning human being seems to be both out of touch and narrow-minded given our modern intuitions about happiness. After all, the average person does not at all seem to delight in theoretical contemplation 24/7. And while there are indeed persons who experience great joy when contemplating and reflecting, there are also persons who lead fulfilling and flourishing lives in other activities. There are those who delight in *contemplating* atop a mountain, and there who those who delight in *climbing* the mountain. Doesn't Aristotle's view seem at odds with the modern view that acknowledges that individuals can lead flourishing lives even if their lives are not dominated by the activity of "doing philosophy"?

In addition to the fact that many fulfilling and happy lives are not dominated by philosophy, the fact that Aristotle claims that there is a single end at which we should all be aimed seems reason enough to reject his theory. Claiming that the "calling" of man is fixed by human nature seems to rest on a theory that is in tension with our modern understanding of human nature. The widely accepted post-modern view of human nature rejects the idea that there are objective constraints on human nature. According to this view, there is no "objective end" that determines how one ought to live in order to flourish. Existence precedes essence, as Sartre says. Flourishing and happiness are not "read off" nature. Human beings create flourishing and happy lives *ex nihilio.* Our modern assumptions about nature, human nature, and flourishing seem to be incompatible with Aristotle's theory.

Flourishing and autonomy: the modern view

The modern view of flourishing and happiness, at least in the West, is one which embraces a robust notion of individuality and individual autonomy. The United States' Declaration of Independence explicitly mentions "life, liberty and the pursuit of happiness" as a right that citizens have against their government. And Supreme Court Justices in the United States have observed that happiness depends on vocational liberty or finding a job that you love to do.[1] The notion that "happiness is finding your passion" is very much a part of modern culture. But on this view, flourishing is largely an *individual* matter. What makes me happy is *not* necessarily what makes you happy. As the saying goes, "different strokes for different folks." Our individualistic notion of flourishing seems to run counter to Aristotle's claim that there is a *single* activity—a specific "calling"—in which flourishing consists. My students often react to Aristotle's ethical theory somewhat contemptuously: *who is he to tell me that I ought to spend most of my time philosophizing?* Justice Stephen Field, who claimed that happiness depends on the liberty to pursue one's own occupation,[2] would surely sympathize with this complaint. Yet, even though the specific activities that comprise a flourishing life may, to a large degree, be an individual matter, Aristotle's theory should not be dismissed so easily.

In this chapter, I begin to construct the argument that shows that the approach to the *NE* argued for in Chapters 2 and 3 of this book can be used to construct a modern theory of flourishing. We can have it both ways: Aristotle's insights about the importance of reason, the commanding nature of affect, and proper pleasure are consistent with a modern view of flourishing in which autonomy is front and center. Far from being irrelevant and outdated, Aristotle's ethical theory can be used to construct a view of human flourishing that is defensible and compelling. I am not alone in believing that Aristotle's approach to ethics remains relevant and important. Daniel Goleman's best selling book, *Emotional Intelligence,* which begins with a chapter entitled "Aristotle's Challenge," acknowledges the importance of cultivated affect or "emotional intelligence" to flourishing. Within philosophy, moreover, "virtue ethics" seeks a return to Aristotle's ethics. In what follows, I show that we can indeed return to Aristotle's ethics without abandoning our belief that autonomy is important to flourishing.

Since my approach fits within the framework of virtue ethics, I begin by discussing the general contours of this theory. After a brief discussion of virtue ethics, I will then discuss an emerging area of research in psychology known as "positive psychology." Those working within virtue ethics seek a view of flourishing that is grounded in the empirical sciences, and positive psychology provides an empirically based account of human flourishing. Within positive psychology, one finds a number of different approaches to flourishing, but I focus on one. This approach, known as "self-determination theory," not only has deep affinities to Aristotle's ethics, but also acknowledges the centrality of autonomy to human flourishing. After discussing self-determination theory, I will discuss some of the important ways in which this modern theory of flourishing incorporates many of the fundamental tenets of Aristotle's ethics. One important similarity, which is the exclusive focus of the following chapter, is that both theories hold that flourishing requires the development and cultivation of specific inner resources. But in addition to this important similarity, there are others just as important, as we shall see.

With this preliminary roadmap out of the way, let's begin by discussing a movement in philosophy that seeks a return to Aristotle's approach to ethics known as *virtue theory*.

Virtue Theory

Moral philosophy has undergone an "Aristotelian turn" due to the work of a number of philosophers such as Phillipa Foot, Elizabeth Anscombe, and Alasdair MacIntyre.[3] Moral philosophy starts with the judgment that some actions are "good" and some are "bad," and the fundamental question for moral philosophy is *in virtue of what?* Do the consequences of an action determine the moral worth of that action? Or is the transgression of some "moral law" the determinant of moral worth? Prior to the work of Foot, Anscombe, and MacIntyre, moral philosophy did not consider flourishing to be relevant to morality. Morality was either about consequences or about rules, not about flourishing. But things have changed. Philosophers have once again taken up the questions "what is the good life?" and "how do I obtain it?" Within moral philosophy, these questions are answered within a paradigm known as "virtue theory" or "virtue ethics." Virtue ethics has also inspired a movement within jurisprudence or legal philosophy known as *virtue jurisprudence*. This approach views the primary purpose of law as the promotion of human flourishing.[4]

Philosophers working within the paradigm of virtue theory presume that morality is *fundamentally* about flourishing. Thus, the project for virtue theorists is to determine the "inner resources" that are characteristic of flourishing lives. According to virtue theorists, the relevant inner resources are intentions, beliefs, and desires. The thought here is that a flourishing life is one that contains *very specific* intentions, beliefs, and desires. Indeed, all individuals have these inner resources. However, not all individuals have flourishing lives. Virtue theorists seek to identify the specific inner resources that are characteristic of a flourishing life. This, of course, raises important questions about the nature of flourishing and about who, among us, is said to flourish. Just as there were competing models of flourishing in Aristotle's day, contemporary virtue theorists likewise have to contend with different models of flourishing.

As we have seen, Aristotle resolves the debate about flourishing by looking to the biological development of human beings. His view is that because our natural, biological development aims at rational activity, a flourishing life is one in which one's calling harmonizes with the target of nature. Thus, a flourishing life is one in which the individual engages in the actions that lead to the cultivation and perfection of rational activity. Today, modern virtue ethicists endorse a strategy similar to Aristotle's. That is, they hold that a conception of human flourishing should be informed by contemporary biology and psychology, both of which shed light on human beings and on human nature. For example, Julia Annas writes:

> Contemporary virtue ethics . . . looks at human nature as we find out about that from the best contemporary science. Here the relevant sciences are biology, ethnology and psychology, studies of humans and other animals as parts of the life on our planet. When we look at other species it has long been clear that we can discern patterns of flourishing particular to the species. There has been reluctance to extend this to humans, on the grounds that we, unlike other animals, can choose and create different patterns of living, and evaluate them, sometimes rejecting and changing them as a result.[5]

As Annas points out, moral philosophy has resisted a naturalistic account of human flourishing because human beings, unlike other species, exhibit great diversity in their life projects and in their political/communal and social organizations. But the fact that there are multitudinous "forms of life" does not mean that we cannot talk about the flourishing of a species whose members have the ability to self-determine. A perfectly reasonable question for human beings is: *given* that a human being can choose his ends and construct a life project—given that I and the members of my species have that unique ability—what is it for me and other members of my species to flourish? Put more starkly, what is human flourishing if, as Sartre says, "man is condemned to be free"? This, according to Annas, is a question that *can* be answered by "the best contemporary science" and that *should* inform moral philosophy.

Virtuous self-determination

Our modern intuitions about flourishing take self-determination seriously. We recognize the variety in life projects and draw from this fact the moral conclusion that human beings *should* be free to construct and determine their own project. This moral conclusion underlies political communities like that of the United States which grants substantial individual rights to citizens. This "bundle of rights" which includes the free exercise of religion, the right to free speech, and the right to privacy, among others, provides a sphere of freedom and autonomy necessary for the creation of one's "life plan" or "personal project."[6] This is a generous sphere of freedom as it presumes that a life project should be one that allows the individual to decide what, if any, religious views he accepts. It also presumes that a life project should be one that allows the individual to decide which trade or trades to pursue. Thus, the bundle of rights enjoyed by those within liberal political communities enables these individuals to determine rather broad contours of their lives. Of course, not all political communities grant to their citizens such a broad zone of autonomy and many human beings have their lives determined for them by forces outside of themselves such as class status. But some political communities do permit their citizens to self-determine, at least to some extent. One important question, therefore, is whether we can speak of flourishing within this sort of political community.

According to Annas and other virtue theorists, we should look to empirical research to help us determine the meaning of flourishing within a political community that allows its citizens to self-determine. As mentioned above, an emerging area of psychology known as "positive psychology" is devoted to the empirical study of human flourishing, and as we shall see, this approach to psychology offers an account of flourishing that is fundamentally Aristotelian and, at the same time, acknowledges the importance of autonomy and self-determination.

Positive Psychology

"Positive psychology" is an umbrella term that encompasses research concerned with flourishing or "optimal functioning" of both individuals and communities. Inspired by the "humanistic" psychological theories of Carl Rogers and Abraham Maslow,[7] positive psychology is a reactionary movement against "the predominant negative bias of traditional psychology."[8] Traditionally, psychologists have "devoted vast attention to the biases, delusions, illusions, foibles, and errors of the human being,"[9] and have not given sufficient attention to the character strengths, virtues, and the conditions that lead to high levels of happiness.[10] Positive psychology seeks to correct this imbalance and to research "the conditions and processes that contribute to the flourishing or optimal functioning of people, groups, and institutions."[11] "Traditional psychology" is focused on detecting and preventing illness and disease, whereas positive psychology is focused on the nurture and cultivation of the traits that conduce to flourishing.

Positive psychology eschews "armchair philosophizing" about human flourishing and instead seeks to understand flourishing by using empirical research methods. For example, positive psychologists typically use survey instruments in which the subject is asked to evaluate life satisfaction and the frequency and intensity of positive emotions. Such surveys measure "subjective well being," or the degree to which the individual experiences positive affect in her life. Another research method used by positive psychologists is known as the "experience sampling method" (ESM). Pioneered by Mihaly Csikszentmihalyi in his groundbreaking research about the phenomenology of optimal experience or *flow*,[12] ESM is a method of recording the different phenomenological experiences that represent a typical week for an individual. Csikszentmihalyi describes this method as follows:

> [The method] involves asking people to wear an electronic paging device for a week and to write down how they feel and what they are thinking about whenever the pager signals. The pager is activated by a radio transmitter about eight times each day, at random intervals. At the end of the week, each respondent provides what amounts to a running record, a written film clip of his or her life, made up of selections from its representative moments.[13]

Csikszentmihalyi's research utilizes a combination of the ESM, surveys, and interviews and is foundational to the positive psychology movement. Taking Csikszentmihalyi's lead, many positive psychologists begin with the premise that flourishing cannot be understood apart from understanding the qualitative differences among the experiences of daily life. As Csikszentmihalyi's research shows, some of these experiences are optimal ("flow" experiences) and some are suboptimal (boredom and anxiety).

The hedonic approach to flourishing

While surveys, instruments and the ESM are used to measure the different affective experiences of which human beings are capable, there is some debate about how subjective experiences should inform our understanding of human flourishing. Is a high degree of positive affect sufficient for flourishing? Or is flourishing something *more* such as a *specific type* of positive affect? Those who believe that positive affect or "subjective well being" is sufficient for flourishing are said to endorse an *hedonic* model of flourishing. On this approach, the focus is on the subjective component of flourishing, that is, pleasurable affect. Normative judgments about these pleasurable experiences are avoided.[14] Essentially, the hedonic approach adopts a view in which flourishing is identified with the subjective experience of maximal pleasure *simpliciter.* On the hedonic approach, therefore, a flourishing life is one in which pleasurable experiences are maximized.

The *eudaimonic* approach to flourishing

Unlike the hedonic approach, the *eudaimonic* approach is substantively normative. That is, it presumes that flourishing consists in "proper pleasures" or "proper functioning."[15] Proper functioning is understood to consist in the satisfaction of *fundamental needs*—needs that are not "up to" an individual. A *eudaimonic* approach distinguishes between merely subjective needs, on the one hand, and fundamental or "objectively valid" needs on the other. The *eudaimonic* approach presumes that flourishing is realized only if fundamental/ objectively valid needs are satisfied.[16] Merely subjective needs produce only "momentary pleasure" when satisfied, whereas objectively valid needs produce truly fulfilling pleasures and hence are essential constituents of flourishing.[17] Abraham Maslow, who theorized that human beings have a hierarchy of needs, endorsed a *eudaimonic* approach to flourishing.[18] Aristotle, of course, adopted a *eudaimonic* approach to flourishing. He rejected simple hedonism and the view that pleasure *simpliciter* ought to be maximized. In his view, man's flourishing lies in his proper functioning, that is, rational activity, and from the specific pleasures that arise from this activity.

Self-Determination Theory

Self-determination theory (SDT) is a *eudaimonic* approach to flourishing that not only has deep affinities to Aristotle's ethics, but can also provide the basis for a modern virtue theory which acknowledges the importance of autonomy.

On the basis of extensive empirical research,[19] SDT researchers have inferred three objectively valid needs that must be satisfied in order for the individual to flourish: the need for *autonomy*, the need for *competence*, and the need for *relatedness*.[20] The need for *autonomy* is the need to feel that one's actions express the authentic choices of the individual; the need for *competence* is the need to feel effective rather than inept in one's environment; and the need for *relatedness* is the need to feel connected to significant others.[21] SDT holds that human beings have "natural innate tendencies" to satisfy these needs.[22] In other words, human beings are intrinsically motivated to engage in actions that are attempts to satisfy these needs.

The exploratory behavior that is ubiquitous among infants, for instance, is taken to indicate an innate tendency for competence in one's environment.[23]

Needs and desires

The fact that SDT adopts a theory of human nature which posits objectively valid needs is important. Given our discussion about the motivational system from Chapter 2, we can understand SDT as envisioning the motivational system as one which is driven by three primary desires. For a need is a desire that has normative force—it is a yearning that *ought* to be fulfilled because such fulfillment is necessary for flourishing. Thus, on the SDT model, the "calling" that drives a human being's motivational system is a complex end that encompasses the three needs of autonomy, competence, and relatedness.

Although Aristotle talks about our yearning for flourishing as a "desire" rather than as a need, the fact that desire is something *normative* for Aristotle makes it akin to the psychological notion of "need." Given that the function of man is rational activity, the desire to know has obvious normative force and hence ought to be fulfilled. Thus, Aristotle claims that we *ought* to desire learning (*NE* 1111a), and that when we learn and know we are brought into our natural condition (*Rhet.* 1371a; *DA* 417b, *NE* 1178a). In the *Politics,* moreover, Aristotle speaks of "doing philosophy" as the only way to satisfy a uniquely human need. In the passage below, Aristotle discusses what motivates a human being to commit criminal acts, and he offers as one possible explanation, the fact that criminals have not satisfied the human need to engage in a specific type of activity that can only be satisfied by engaging in philosophical activity:

> [W]ant is not the sole incentive to crime; men also wish to enjoy themselves and not to be in a state of desire—they wish to cure some desire, going beyond the necessities of life, which preys upon them; nay, this is not the only reason—they may desire superfluities in order to enjoy pleasures unaccompanied with pain, and therefore they commit crimes. Now what is the cure of these three disorders? Of the first, moderate possessions and occupation; of

> the second, habits of temperance; as to the third, if any desire pleasures which depend on themselves, they will find the satisfaction of their desires nowhere but in philosophy; for all other pleasures we are dependent on others. (*Pol.* 1267a)

In this passage, Aristotle alludes to both a need to cure our "state of desire," as well as a need to engage in very specific kind of activity—a pleasurable activity that does not depend on others. And as he indicates, only philosophical activity can satisfy this need. Clearly, Aristotle's theory is one that easily accommodates the psychological notion of a need. It is quite reasonable to interpret Aristotle's view as one which endorses the idea that human beings have a need for rational activity.

In elucidating further details about the affinities between SDT and Aristotle's ethics, I am going to narrow my discussion. As we have seen, SDT envisions three needs at the core of human nature: the need for autonomy, for competence, and for relatedness. I will focus the remainder of my discussion on just one of these needs—the need for autonomy. The other needs recognized by SDT—competence and relatedness—clearly have affinities with Aristotle's view. Aristotle's emphasis on knowledge and contemplation speaks to *competence*, and his discussion about the important role that friendship plays in the good life (*NE* 1155a–1172a) speaks to *relatedness*. But autonomy is central to our modern view of flourishing and is seemingly absent from Aristotle's theory. Since one of the principal aims of this book is to show how Aristotle's theory is compelling and relevant to a modern view of flourishing in which autonomy is central, the need for autonomy deserves special attention.

The need for autonomy

According to SDT, autonomy is a fundamental psychological need—an innate yearning that creates disequilibrium if unsatisfied. How is the need for autonomy satisfied? Importantly, the need for autonomy is fully satisfied only if the individual engages in certain type of activity, specifically, a certain type of goal-oriented or teleological activity. The classical liberal account of autonomy identifies it simply as the "absence of constraints."[24] But SDT rejects this view of autonomy. Instead, SDT conceives of autonomy as the *presence* of a special kind of teleological experience. SDT identifies autonomy as a goal-oriented experience, albeit one of a very specific kind.

Varieties of teleological experience

In order to understand the nature of the goal-oriented activity with which SDT identifies autonomy, we need to distinguish between two very different ways in which an individual can experience her teleological or goal-oriented activity. Goal-oriented or teleological behaviors have distinct "phenomenological signatures"—flavors, really—depending on the *perceived origin* of these behaviors. One can experience one's teleological behavior as emanating from two very distinct causes, and each experience has a distinct "qualitative feel" for the individual engaging in it.

To illustrate, suppose Paul decides to read Homer's *Iliad*. And further suppose that Paul's decision to read this book has nothing to do with an assignment for a class. Rather, he decides to read this book because he loves Homer. Paul is drawn—indeed compelled—to read the *Iliad* because the activity of reading this type of work is pleasurable for him. In this example, Paul reads the *Iliad* for its own sake, and his reading of this book springs from *intrinsic motivation*. Because Paul knows that this activity comes from his own preferences and desires, he has a specific perception of himself when he engages in it. Specifically, Paul perceives his reading of the *Iliad* as emanating from what is referred to as an "internal perceived locus of causality." It is *Paul's* desires and interests that are the origin of his act of reading the *Iliad*. When one perceives one's teleological behavior in this way, one sees one's behavior as issuing from the "core self." So, as he reads the *Iliad*, Paul sees himself engaging in an activity that originates and issues from *him*.

However, suppose Paul hates reading biographies and any work of non-fiction. Further suppose that he is assigned to read Hillary Rodham Clinton's *Living History*. Given that he hates reading biographies and non-fictional works, he will be, as we say, "un-enthused" to engage in this task. He will not find reading the book enjoyable or pleasurable. Paul would never read this book for its own sake. If he does read it, he will do so because he "has to" as this is his assignment. His behavior will therefore be *extrinsically motivated.* Paul's reading of *Living History* does not issue from him, but from the teacher who assigned it. Therefore, as he reads this book, he perceives himself to be engaged in an activity that originates and issues from another. When one perceives that one's teleological behavior issues from something outside of oneself such as an authority figure, then one's perception proceeds from an "external perceived locus of causality."

Intrinsic and extrinsic motivation

The distinction between internal and external perceived loci of causality acknowledges the distinction between the different motivational forces that drive our goal-oriented, teleological behavior. Our purposive behavior can be "intrinsically motivated" and hence driven by innate drives and preferences, or it can be "extrinsically motivated" and hence driven by forces outside of ourselves. Intrinsically motivated actions "come from us," whereas extrinsically motivated actions do not. Granted, there is a sense in which all actions "come from us." Consider: even when Paul reads *Living History* solely for the sake of completing an assignment, his decision to do so "comes from him" since he forms the intention to perform this task. He finds a quiet room to work and says "ok, I need to read Clinton's autobiography. I need to do my homework" and, conscientious as he is, devotes several hours to the task. No one is *forcing* Paul to do this so, in a sense, his action comes from him. However, there is a sense in which the action *does not* come from him. Because he hates biographies, he would never choose to read this book for its own sake. The act of reading *Living History* is, therefore, called into existence by a force that is *outside* of Paul (his professor). In this way, his act of reading *Living History* is caused by something external to him.

Self-concordant activity

Researchers have a special name for goals that originate from, and hence provide intrinsic motivation for, the individual pursuing them.

Such goals are called *self-concordant goals*. Thus, self-concordant activity is intrinsically motivated teleological activity. Research demonstrates that those who engage in self-concordant activities exhibit a higher degree of well-being than do individuals who tend to engage in activities for "external" reasons.[25] Individuals who pursue self-concordant activities demonstrate greater creativity, increased cognitive skills, greater self-esteem, and satisfaction.[26] Some research suggests that self-concordance correlates with well-being irrespective of cultural membership.[27] Even individuals who belong to more "communitarian" cultures such as that of China experience greater well-being if they are self-concordant.[28]

Importantly, self-concordant activity is not merely *any* pleasurable activity. Rather, a self-concordant activity is one which aligns with an individual's deep values and interests.[29] For instance, if generosity and concern for others are important values to an individual, and if the individual is interested in world music,[30] a self-concordant activity may be volunteering to teach area youth about world music. Self-concordant activities can also be the basis of one's professional work. For example, the activity of teaching and learning about biology might be a self-concordant activity if one deeply values knowledge and has an enduring interest in biology. Differences in personality will yield different value hierarchies, but the recipe for a self-concordant activity is the same: an activity that expresses the core values and interests of an individual.[31] Self-concordant activities are said to "emanate from one self" whereas "external activities" are "brought about by pressures and forces external to the self."[32]

The self-perception of autonomy

We have been discussing self-concordant activity as the specific type of activity by which one satisfies their fundamental need for autonomy. Self-concordant activity does this because it provides a very specific phenomenological experience to the individual engaging in it. Self-concordant activity allows the individual to perceive one's actions as issuing from oneself. Thus, self-concordant activity is important not only because it is intrinsically motivated and meaningful, but also because it provides a particular experience of an individual's own agency and teleological behavior. Self-concordant activities allow one to experience one's actions as following from oneself and hence as authentic.

The pleasures of self-concordance

Because self-concordant activity is intrinsically motivated activity, it provides intensely pleasurable affect for the individual engaging in it. Returning to our example of Paul reading Homer's *Iliad,* we should point out that Paul's activity of reading the *Iliad* yields a *fusion* of pleasures that give self-concordant activity its particular "hedonic signature." Not only is the act of reading the *Iliad* a source of pleasure (Paul loves reading Homer), but the perception of his actions as intrinsically motivated is *also* source of pleasure. Thus self-concordant activity provides a unique and intense pleasure for the individual engaging in it: not only does the individual experience the pleasure and enjoyment from the activity that aligns with her values and interests, but she also experiences the pleasure of her own agency—of seeing her actions emanate from herself rather than from another. Self-concordant activity provides a potent "hedonic cocktail" that quenches the individual's innate thirst for autonomy.

Thus far, we have seen that SDT claims that autonomy—understood as self-concordant activity—is a necessary component of a flourishing life. At this stage, let's note some of the deep affinities between this tenet of SDT and Aristotle's view of flourishing.

Self-Determination Theory and Aristotle's Ethics

Aristotle on intrinsic motivation

One important similarity between the SDT model of flourishing and Aristotle's, concerns the centrality of intrinsically motivated activity. As we have learned, SDT claims that intrinsically motivated activity is essential to flourishing. Aristotle shares this insight. Indeed, intrinsic motivation is central to Aristotle's account of flourishing.

The notion of intrinsically and instrumentally desirable goods plays a fundamental role in Aristotle's definition of flourishing. Typically, however, the phrase "intrinsically desirable good," evokes an image of a good *qua thing*. When we hear the word "good," our tendency is to sever it from the motivational system that pursues it. We focus on the good as an *object* that possesses the "good in itself" attribute, and we ignore the motivational system that pursues it. However, it is important to remember that for Aristotle, intrinsically desirable goods can be fully understood only in relation to motivation and the motivational system. The notion of a "good" for Aristotle is necessarily connected to *action*. A good is a *final cause*—a goal—of action and the chief good is the *most final cause* of action (*NE* 1094a–b). Since a "good" is necessarily connected to action, then so are intrinsic and instrumental goods. One cannot understand the notion of an "intrinsically desirable good" independent of intrinsically motivated activity. One implies the other.

To conceive of an "intrinsically desirable good" as a tangible entity that, perhaps, sits atop some objective normative hierarchy is to miss the fact that Aristotle's notion of good is necessarily connected to action and hence to *desire*. Aristotle, recall, believes that actions proceed from desire (*DA* 433a). Thus a "good" is an *object of desire* that stimulates action for its sake, and objects of desire have this power because they ultimately lead the individual to something *pleasurable*.

As we know, an object of desire need not be a thing, but can also be an activity. Aristotle, recall, regards activities as sources of pleasure (*NE* 1175a). In fact, he says that activities are *necessarily connected* to pleasure:

> [T]he pleasures involved in activities are more proper to them than the desires; for the latter are separated both in time and in nature, while the former are close to the activities, and so hard to distinguish from them that it admits of dispute whether the activity is not the same as the pleasure. (*NE* 1175b)

Aristotle is pointing out that certain activities are sources of pleasure. Indeed, as he notes in the last sentence, the various pleasures are, for all intents and purposes, *different activities*. Pleasure, however, is an intrinsic motivator. As Aristotle observes, "no one asks to what end one is pleased" (*NE* 1172b). Given that pleasures are necessarily yoked to activities, it follows that activities are intrinsic motivators and hence function as "goods." An activity can exert a "magnetic pull" on us so that we are compelled to perform actions for its sake, and for its sake alone. Activities provide intrinsic motivation for the individual to pursue them because they are pleasurable.

We noted earlier that Aristotle claims that some activities are the *right* intrinsic motivators, whereas other activities are the wrong ones:

> [T]o each activity there is a proper pleasure. The pleasure proper to a worthy activity is good and that proper to an unworthy activity bad; just as the appetites for noble objects are laudable, those for base objects culpable. (*NE* 1175b)

Not all people pursue the same activities. Whether a particular activity is an intrinsic motivator varies from person to person (*NE* 1153b). Moreover, most people pursue the *wrong* activities (*NE* 1095b). Why? Because, as Aristotle says, they have not truly tasted the superior pleasures associated with rational activity (*NE* 1176b). Aristotle's discussion about the "real" and the "apparent" good (*NE* 1113a) is related to this claim. Those who are commanded by the "apparent" good are indeed intrinsically motivated, but they are intrinsically motivated to pursue the *wrong* pleasurable activities. Thus, the affective content of the calling that drives the motivational system of such individuals is *wrong*. However, those who are commanded by the real or true good are intrinsically motivated to pursue the right pleasurable activities. The affective content of the calling that drives *these* individuals is correct.

Clearly, the notion of intrinsic motivation is central to Aristotle's ethics. To flourish is to be intrinsically motivated to pursue the right pleasurable activities and hence *rational* activities. And as we noted in the last chapter, the default state of the motivational system is *not* one in which the individual is intrinsically motivated to pursue the right activities. Rather, the default motivational system of children is intrinsically motivated to pursue activities that yield *bodily* pleasures (*NE* 1154b, *NE* 1179b, *NE* 1111a, *NE* 1111b). Only with a certain kind of training and cultivation can one awaken the proper pleasures so as to alter the nature of the activities that the individual is intrinsically motivated to pursue. And once that happens, the individual is intrinsically motivated to pursue the pleasures that are yoked to *rational* activity. The falcon can hear the falconer.[33]

Essentially, then, the aim of habituation and training is to change the nature of the activities that the individual is intrinsically motivated to pursue. This is evidenced by the fact that Aristotle states that one is not virtuous *unless* one is intrinsically motivated to perform virtuous actions (*NE* 1095a). As he says, in order for an act to be virtuous, (1) the individual must know what he is doing, (2) he must choose the action for its own sake, and (3) the action must spring from an established habit (*NE* 1105a). Aristotle clearly states that an intrinsic motivation to perform virtuous actions is a *necessary condition* of virtuous behavior. And interestingly, he says that the knowledge condition "counts for little or has no weight" (*NE* 1105b). What matters for virtuous behavior is *not*, therefore, that the individual *knows* that he is behaving virtuously. Rather, the important thing is that the individual is intrinsically motivated to behave virtuously.

Aristotle's theory of flourishing is one in which the notion of intrinsically motivated activity is central.

Aristotle on the pleasure of autonomy

In addition to the fact that both SDT and Aristotle claim that flourishing requires intrinsically motivated activity, another similarity between the two theories concerns the importance of autonomy. Although the notions of autonomy and self-concordance may seem to be a far cry from Aristotle's philosophy, the fact is, in a number of passages, Aristotle indicates the important role of voluntary, non-coerced actions. And interestingly, in the *NE*, he notes that a certain type of self-perception, the self-perception that one is engaged in virtuous activity, is its own source of pleasure:

> But if life itself is good and pleasant (which it seems to be, from the very fact that all men desire it, and particularly those who are good and supremely happy; for to such men life is most desirable, and their existence is the most supremely happy) and if he who sees perceives that he sees, and he who hears, that he hears, and he who walks, that he walks, and in the case of all other activities similarly

> there is something which perceives that we are active, so that if we perceive, we perceive that we perceive, and if we think, that we think; and if to perceive that we perceive or think is to perceive that we exist (for existence was defined as perceiving or thinking); and if perceiving that one lives is in itself one of the things that are pleasant (for life is by nature good, and to perceive what is good present in oneself is pleasant); and if life is desirable, and particularly so for good men, because to them existence is good and pleasant for they are pleased at the consciousness of the presence in them of what is in itself good); and if as the virtuous man is to himself, he is to his friend also (for his friend is another self):—if all this be true, as his own being is desirable for each man . . . because he perceived his own goodness, and such perception is pleasant in itself. (*NE* 1170a–b).

In this passage, Aristotle claims that the self-perception that one is living is pleasant. Indeed, his claim is that the self-perception that one is living virtuously is even more so. Doubtless you have experienced the pleasure upon performing some action that you know "is right." Aristotle is alluding to that type of experience in the passage above.

The above passage occurs within Aristotle's discussion of friendship, and Aristotle's remarks are made in an effort to show that, since the virtuous man takes pleasure in the perception that he lives and acts with virtue, then he will also take pleasure in seeing *his friend* live and act with virtue. And although, in the above passage, Aristotle does not explicitly state that good men take pleasure in seeing themselves act autonomously, elsewhere, he does acknowledge that voluntary actions, actions which issue from oneself, are pleasant.

> Moreover, as all actions due to ourselves are done voluntarily and actions not due to ourselves are done involuntarily, it follows that all voluntary actions must either be or seem to be either good or pleasant. (*Rhet.* 1370a)

It follows from these two claims that the self-perception of voluntary action is a very specific type of pleasurable activity. For both the performance of the voluntary action *and* the perception of oneself performing the voluntary action are sources of pleasure.

Recall, though, that Aristotle recognizes two different types of voluntary actions. There are some forms of voluntary action in which children and animals partake (*NE* 1111a), but man is capable of a different type of voluntary action which involves the direction of practical reason (*NE* 1102a–b). Now consider the self-perception of one engaging in *correct* rational voluntary action—action directed by the correct rule. Let's refer to this latter type of voluntary action as *rational* voluntary action. Now consider the self-perception of one engaging in rational voluntary action. It follows from Aristotle's claims above that one who perceives oneself acting pursuant to the correct rule will have a very distinct pleasurable experience. In other words, the perception of oneself engaging in correct rational voluntary activity has a specific "hedonic signature," and serves up a rather potent "hedonic cocktail."

When we consider Aristotle's remarks about *contemplation,* moreover, it becomes clear that, one of the reasons that contemplative activity offers the highest and best pleasure for man, is due, in part, to the fact that it is rational voluntary activity *par excellence.* As we noted earlier, contemplative activity is pleasurable because it is a rational activity, and, as such, *must* be pleasurable to a human being who has cultivated affect. But contemplative activity is valuable for another reason. It is the only activity that can be performed truly autonomously:

> And the self-sufficiency that is spoken of must belong most to contemplative activity. For while a philosopher, as well as a just man or one possessing any other virtue, needs the necessaries of life, when they are sufficiently equipped with things of that sort the just man needs people towards whom and with whom he shall act justly, and the temperate man, the brave man, and each of the others is in the same case, but the philosopher, even when by himself, can contemplate truth, and the better the wiser he is; he can perhaps do so better if he has fellow-workers, but still he is the most self-sufficient. (*NE* 1177a)

Contemplative activity is integral to flourishing not only because it is rational, but also because it is "self-sufficient." But why should this matter for Aristotle? Why, after all, does it matter to Aristotle that the specific activity in which flourishing consists is "self-sufficient"?

As we know from Chapter 1, Aristotle claims that the specific activity in which flourishing consists must be something that is unique to human beings (*NE* 1098a). And we noted above that the self-perception of rational voluntary activity provides a pleasurable experience unique to human beings. Now, contemplative activity is a *special kind* of rational voluntary activity—one that can be engaged in even when one is alone. Contemplative activity requires no props. Let's call *this* kind of rational activity *super* rational voluntary activity. The following table summarizes the different kinds of voluntary activity.

Varieties of voluntary activity

	Description	**Example**	**Source of pleasure?**	**Pleasure unique to human beings?**
Voluntary activity	The source of activity originates from the animal.	Horse leading himself to water. Baby crawls towards toy.	Yes	No
Rational voluntary activity	The source of activity originates from the individual listening to the correct rule apprehended by practical reason.	Archie Gates listens to the rule that says "you must act courageously" and faces his enemy in battle.	Yes	Yes
Super rational voluntary activity	The source of activity originates from the individual listening to the correct rule apprehended by practical reason AND the individual can engage in this activity on her own. Activity requires no props.	Contemplation	Yes	Yes[34]

Because contemplation is the only kind of super rational voluntary activity, the self-perception of one engaging in this activity will therefore give rise to a very distinct human pleasure. Indeed, contemplative activity is the *only way* by which one can experience this pleasure. Such a pleasure is associated with a rich and robust perception of one's voluntariness—one's own agency—since one can engage in contemplative activity alone, and with no props. In the passage below, which occurs within Aristotle's discussion about why some individuals engage in criminal activity, he indicates that human beings have a *need* to experience the pleasure arising from the performance of an activity that depends on nothing but oneself:

> [I]f any desire pleasures which depend on themselves, they will find the satisfaction of their desires nowhere but in philosophy; for all other pleasures we are dependent on others. (*Pol.* 1267a)

Contemplative activity is the only rational activity that can be engaged in without any resources other than the self. Indeed, in this passage, Aristotle appears to agree with SDT that human beings have a need to engage in pleasurable activity that is "their own," and to see themselves doing so. Given that the self-perception of rational voluntary activity is special type of pleasurable activity available to human beings, and given that contemplative activity is rational voluntary activity *par excellence,* then contemplative activity is *naturally* very pleasurable for a human being. Contemplative activity consists in two uniquely human pleasures: the pleasure of rational voluntary activity, on the one hand, and the pleasure arising from the *self-perception* of this special rational voluntary action, on the other. No wonder Aristotle claimed that contemplative activity is the highest and best activity in which human beings can engage.

The important point is that Aristotle's focus on the importance of self-sufficiency reveals that he shares in the opinion of SDT that the self-perception of one's action as voluntary—as one's own—produces a unique human pleasure that is a fundamental component of human flourishing.

The ownership of one's virtue

Another indication that Aristotle valued one's self-perception of autonomy can be gleaned from his remarks about responsibility. Aristotle claims that virtue is a matter of individual responsibility and is "up to us" (*NE* 1113b). And it is significant that Aristotle says this given that the students of the *NE already* have good habits (NE 1095b). Despite the students' excellent training, Aristotle nevertheless feels it important to make explicit that the acquisition of virtue is fundamentally an act of individual will. Given that the *NE* is intended to be a transformational work, and hence that Aristotle aims to make the students good, then Aristotle must make this remark because he expects it to have a certain effect on his pupils. In highlighting the students' responsibility, Aristotle seems to acknowledge the importance of explicitly recognizing his students' agency so that the students perceive their virtuous actions as "their own." The question, however, is *why* does Aristotle believe that having his students understand that they, ultimately, are responsible for their virtue, will contribute to the transformative aim of the *NE*?

In Chapter 2, I argued that Aristotle's aim in the *NE* is to promote the moral development of his students by exposing them to a different type of rational activity, a self-reflective activity in which the student considers his own virtuous actions as a starting point for inquiry. But when we consider the fact that Aristotle places importance on rational voluntary actions, we can also understand that the lessons of the *NE* have yet *another* kind of impact on the student. Let me explain.

The effects of reflecting on one's virtue

The fact that the students are responsible for their training does not imply that they *perceive* their virtuous behavior as "their own." Recall our above example in which Paul reads the *Living History* because he is assigned to do so. Suppose he begrudgingly complies. Crucially, Paul's decision to do the assignment originates *in him*, but he *does not* perceive his act of reading to be an authentic, intrinsically motivated act since the act is prompted by his professor. So why are Aristotle's students any different? Sure, the decision to become virtuous "rests with them" but perhaps they do not see their virtuous behavior as "their own." The students, so the argument goes, are "conformers." They have submitted to and accepted their community's standards of excellence. Indeed, these external standards have guided their practice and training. Since these norms and standards are "outside" and external to the students, then why should they perceive their virtuous behavior as "their own"?

Autonomy and habituation

In responding to this question it is important to point out that the behavior of the students is habitual (*NE* 1105a). Such behavior does not occur for an extrinsic reason, such as avoiding the punishment associated with non-compliance. It occurs automatically, habitually. Given this fact, shouldn't the students perceive this behavior as authentic? Actually, not necessarily. The mere fact that a behavior is performed habitually indicates nothing about whether the individual perceives this behavior as authentic. Consider, for instance, a smoker who desperately wants to quit, but cannot break the habit. Surely, this individual does not regard her unwanted behavior as emanating from her core self. Nor does an individual regard her habit of stopping at a red light as an authentic act. So, although the students' behavior is habitual, it does not follow that they perceive this behavior as authentic.

Primitive and normative habits

Yet, the fact that a behavior is habitual does not *preclude* it from being perceived as one's own. To see this, let's make some distinctions among different kinds of habitual behaviors. Consider the differences between the habitual behaviors mentioned above: *smoking* and *stopping at a red light.* The former does not result from any sort of training—that is, some external authority imposing rewards or punishments so as to mold behavior to conform to some external norm or standard. Rather, the habit of smoking *emerges* from the individual's voluntary behavior, without external prompts.[35] Because this sort of habit is not the product of training aimed at compliance with an external norm or standard, let's refer to it as a *primitive* habit. Children's habits such as thumb sucking are primitive habits because they emerge from the child's own impulses. In contrast to primitive habits are *normative* habits. Normative habits do not emerge from one's un-coerced actions and impulses. Rather, they originate from repeated compliance to some external norm or standard. Normative habits are automatic, non-deliberative behaviors which result from systematic training aimed at compliance with external norms. Stopping at a red light, using the toilet, and chewing with one's mouth closed are all examples of normative habits. We just do these things automatically, but we had to be trained to comply with external norms to get there.

Internalization and integration of norms

Because the students' virtuous behavior embodies their community's norms of excellence, they have normative habits. The question, then, is this: can one *ever* perceive normative habits as "one's own"? And if so, do the students of the *NE* do so?

Self-determination theory helps to construct an explanation of how normative habits can be perceived to be authentic and autonomous. According to SDT, if one *integrates* the external norms from which one's normative habits flow, then she can experience such habits as authentic.[36] *Integration* of a norm occurs when the individual has consciously evaluated the norm and has assimilated it into her core values.[37] Integration is contrasted with, and presupposes, *internalization*. An individual *internalizes* a norm when her compliance becomes "natural" and hence external punishments/rewards are no longer necessary to engender such compliance. Internalization is really just habituation, albeit habituation that lacks reflection and integration.

To further illustrate what is meant by internalization and integration, recall Paul from our earlier discussion. When he completes his assignment "read Clinton's *Living History*," he complies with the external norm *one ought to do one's homework assignments*. Now let's suppose that Paul's compliance with his norm is habitual. He just does it. The important question is *why*, what is motivating Paul to comply with this norm? One possibility is that Paul habitually complies with this norm merely because of the guilt that he feels from not complying. If so, then Paul has internalized the norm—he no longer needs the external prompts to mold him into compliance. But suppose Paul habitually complies with the norm because he understands that complying with this norm helps him to develop skills that will be useful to his self-concordant goal of being a teacher. If this is the source of compliance, then Paul has assimilated this external norm into his core values and hence has *integrated normative habits*. When this happens, Paul perceives his normative habits as "his own."[38]

Ultimately, whether normative habits are perceived as "one's own" depends upon whether the individual has consciously and deliberately integrated the norms from which this behavior springs. Thus, the mere fact that one's habitual behavior embodies external norms does not preclude this behavior from being perceived as "one's own." Consider the numerous individuals who have a passion in life for activities that are governed by rules (chess, soccer, etc.). David Beckham, whose passion is soccer, engages in numerous normative habits. The rules of soccer are "external" to him, but his compliance with these rules is habitual and automatic. Because he is passionate about soccer, and therefore has chosen this activity as his end and his calling, he perceives this behavior as "his own." Thus it is possible to engage in habitual normative behavior that is perceived to be authentic. Anyone who has a self-concordant end that includes some rule-governed activity will attest to this fact.

To summarize: there are three types of habits: (1) primitive habits which do not result from training aimed at compliance with an external norm; (2) normative habits which are the result of training aimed at compliance with external norms, but which lack reflection; and (3) integrated normative habits which are normative habits upon which the individual has reflected and consequently assimilated into his core values, thereby accepting them as his own:

Primitive habits	Habitual behavior that does not result from training aimed at compliance with external norms (e.g., nail biting, hair twisting, internet surfing).
Normative habits	Habitual behavior that results from training aimed at compliance with external norms (e.g., stopping at a red light; using the toilet; brushing teeth). All normative habits proceed from internalized norms.
Integrated normative habits	Normative habits that proceed from integrated norms—i.e., external norms that have been reflected upon and subsequently assimilated into the individual's core values by the individual himself (e.g., Paul does his homework assignments because he sees them as necessary to realizing his self-concordant end of becoming a teacher).

Now that we understand the different varieties of habitual behavior, let us return to the students of the *NE*. At the very least, they have normative habits. They have internalized their community's norms. We know that the students choose virtuous behavior for its own sake, and hence no longer need the external rewards/punishments to induce compliance.

There is no evidence, however, that Aristotle's students have *integrated* their community's norms prior to the lessons of the *NE*. In all likelihood, they have *not* engaged in the self-reflection necessary for integration and hence have not considered why compliance to their community's norms is important. But of course, the students *could* engage in reflection about these norms. And the *NE* could reasonably be interpreted as a tool that helps the students to do this.

Recall that the *NE* engages the student in an act of self-reflection. Again, the students' own actions provide the starting point of inquiry. As Aristotle says, the subject matter of the *NE* is, essentially, the actions of its pupils (*NE* 1095a). Thus, through these lectures the pupils learn about the role that their own normative habits play in their own flourishing. As students of the *NE*, Aristotle's pupils can begin to understand the *why* of their virtuous habits, and *how* they contribute to their own flourishing. And one could reasonably interpret the *Politics* as an effort to deepen this reflection. The *NE* ends by telling the student, *now we need to study different normative ecologies*, which is a segue into the *Politics*. The *Politics* involves an assessment of the degree to which different normative ecologies (moral customs, laws & political constitutions) achieve the objective of human flourishing. Thus, a student who undertakes an evaluation of these normative ecologies will be involved in reflection about how his own normative ecology fares in comparison with others. If such reflection results in full integration, the student could thereby perceive his virtuous habits as fully authentic and as "his own."

Of course, reflection that involves comparative theorizing could also cause the student to realize that his own normative ecology could be improved. Such a student could then undertake action aimed at its improvement. Since the student is already habitually virtuous (*NE* 1095b), his endeavor to improve his community's normative ecology will express affective virtue. In other words, the actions undertaken to improve his normative ecology will proceed not from the bodily passions, but from a motivational system driven by the proper, rational pleasures. Put more colloquially, the student's desire to improve his community will be motivated by the "right heart" or the "correct calling."

The phenomenology of habitual action may include a sense of ownership and authorship, or it may not. The transformative aim of the *NE* considered in conjunction with the importance that Aristotle places on voluntary action, makes it reasonable to think that Aristotle believed it was important for his students to reflect on their normative habits in order to deepen their perception that their behavior is truly voluntary. Again, a student's perception of his behavior as truly voluntary can be achieved if 1) he integrates his community's norms and therefore sees them as essential to his own flourishing; and/or if 2) he engages in actions to improve his normative ecology so that it is better able to promote human flourishing. In either case, the reflective activity stimulated by the *NE* and the *Politics* can profoundly deepen the extent to which the student perceives his actions as his own.

The fact that the *NE* engages the student in self-reflection, leads to the reasonable conclusion that these lectures had a powerful effect on the student's perception of the norms that are at the basis of his habits. For by reflecting on these norms—and on the ecology of which they are a part—the student is able to integrate them and/or to act to change them so that they better realize human flourishing. Such reflection and/or action allows the student to see his habitual behavior as truly voluntary. The difference between internalized and integrated norms helps us to see that Aristotle may have very well intended his lectures to provide the student with a deeper appreciation of his community's norms, a deeper sense of the ownership of his actions, and hence a more robust perception of his own agency.

Clearly, there are deep affinities between the theories of flourishing offered by SDT and Aristotle. Because SDT is a eudaimonic theory of flourishing, it shares in Aristotle's belief that flourishing consists in activity that is motivated by the correct affective content. Moreover, both theories acknowledge that a special type of intrinsically motivated teleological activity is essential to flourishing. Both theories also recognize the unique pleasure arising from voluntary activity and the self-perception thereof. Some of Aristotle's remarks can reasonably be interpreted as an endorsement that human beings have a *need* to engage in authentic, voluntary activity, and to experience the pleasure associated with the uniquely human experience of being able to perceive oneself engaging in such activity.

But what of the notion of virtue? Virtue, after all, is at the heart of Aristotle's theory. Does SDT and the notion of "self-concordance" enable us to develop a modern account of virtue? Indeed it does.

Aristotle tells us that a virtue is what makes the work or *ergon* of a human being be done well (*NE* 1106a). For Aristotle, this *ergon* is rational activity. As we discussed in Chapter 2, contemplative activity, the highest and best rational activity, is possible only if one possesses the "inner resources," that is, the virtues, on which it depends. Contemplative activity is a *masterful* rational activity that is possible only if the individual possesses a cultivated and virtuous motivational system. The possession of such a motivational system is a developed capacity that, as we have seen, begins with habituation. Habituation not only awakens an individual to his proper pleasure, but it also provides excellent epistemic skills that are necessary for knowledge and wisdom.

In the following chapter, I show how the fundamentals of this story remain the same if we adopt a view in which flourishing is identified with self-concordant activity. Self-concordant activity is a masterful activity that is possible only if one possesses the "inner resources" and virtues on which it depends. To understand the nature and development of these virtues, we will need to shift focus to an area of empirical research that is focused on the attainment of goals. This research will not only help us to better understand the preconditions for self-concordance, but it will also provide us with materials with which to construct a modern virtue theory. Let us now turn to a discussion about the inner resources that are necessary for self-concordance.

Chapter 5

The Developmental Preconditions of Self-Concordance

Introduction

In discussing the various competencies underlying self-concordance, I appeal to some of the scientific literature on goal attainment. As this literature is discussed, remember that self-concordance is a species of goal-oriented action. True, self-concordant activity is a *special kind* of goal-oriented action involving intrinsic motivation and the enduring values and interests of the individual, but "stripped bare" of these accoutrements, at the base of self-concordant activity is the basic ability to set and realize a goal. It is this basic ability on which I will focus.

As we shall see, the ability to deliberately set and realize a goal is far from simple but involves a variety of components that must be developed and nurtured. In order to see this, we need to turn to an important concept that is found within neuroscience and developmental and educational psychology, that is, *executive function.* It is executive function that underlies a human being's ability to engage in goal-oriented activity. Practical reason, recall, is the specific rational ability earmarked for goal-oriented activity. So, as we discuss the research on the cognitive and affective components involved in goal attainment, keep in mind that these are the components underlying *practical reason.* Now in developmental psychology, the term "executive function" is used instead of the term "practical reason." However, the terms are, for all intents and purposes, interchangeable.

Executive Function

Executive function is an important construct that drives empirical research in neuroscience and developmental and educational psychology.[1] The amount of research on this construct is voluminous and spans topics from attention deficit hyperactivity disorder (ADHD) to working memory.

Executive function is important because it provides an empirical correlate of the concept of practical reason. Executive function includes a suite of abilities that are involved in ongoing goal-directed behavior. Some of these abilities consist in the "analytic" and intellectual abilities involved in goal-oriented behavior. For example, executive function is responsible for "surveying information, imposing organization, formulating plans, setting goals, keeping track of changing circumstances, and even anticipating numerous possibilities and modifying goals and plans accordingly."[2] However, executive function also includes affective strengths such as inhibition and self-regulation.

For the sake of perspicuity, I divide the executive function into three different components: the intellectual, the navigational, and the motivational. Some narrow definitions of executive function could very well exclude the motivational component that I discuss. Narrow definitions of executive function are those that focus on the cognitive components of goal-directed behavior (such as planning) rather than on the somatic or bodily components that are necessary to implement the plan. But since a superb planning ability is not sufficient for realizing a goal (one could fail to implement the plan), and since we are concerned with the underlying components involved in the realization of a goal (since self-concordance is a specific type of goal realization), I adopt a broad definition of executive function that includes motivational competencies.

It is worth noting that the suite of abilities comprising the executive function resides in the dorsolateral prefrontal cortex of the brain (the area behind the forehead).[3] Individuals with damage to this part of the brain (as well as individuals with Parkinson's disease) are found to have impaired teleological abilities.[4] One famous example, discussed by Antonio Damasio in his book *Descartes' Error*, concerns the case of

Phineas Gage. After Gage suffered accidental damage to the frontal cortex, his personality changed drastically. Prior to the accident, Gage had been a responsible and hard-working individual. After the accident, Gage had lost his ability to control base impulses. He became impatient and was unable to follow through with plans.[5]

The prefrontal cortex is the last part of the brain to develop and does not reach maturation until the 20s.[6] Research in child development reveals that both the preschool years[7] and the adolescent years are crucial to the proper development of the executive function.[8] These facts are of course interesting given Aristotle's claim that rational activity comes last in the development of a human being (*GA* 736b). Developmental neuroscience reveals that Aristotle was correct about this. The part of the brain responsible for rational voluntary activity is formed last in the brain's natural development.

With these preliminary remarks about executive function out of the way, let's begin by discussing the intellectual component of the executive function.

The Intellectual Component of Executive Function

The intellectual component of the executive function is simply a type of "strategic planning" or "practical deliberation" by which one determines the orderly sequence of actions required to bring about some end. The practical knack involved in realizing long-term goals requires the ability to "think backwards," that is, to reason from end to means and therein to derive the specific sequence of steps required. Think of this ability as the planning ability of executive function.

We noted earlier that Aristotle describes practical reason as a distinct form of rational activity that deals with the contingent, the uncertain, and the variable (*NE* 1143a). The epistemic skills involved in practical reason include the ability to chart an effective course that acknowledges the dynamic, unpredictable nature of human life. Since practical reason operates within a domain of circumstance and contingency, a component of executive function known as "set-shifting" enables one to use feedback from the environment in order to modify one's plan so that one "stays on target" relative to one's goal.[9] For instance, if one's goal is receive an "A" in a class and the outcome of a particular study strategy is a "C," the set-shifting ability underlies one's ability to formulate an alternative strategy in light of this information. Such an ability requires that the individual can shift attention from one strategy to another upon receiving feedback that the first strategy is ineffective. Attention and memory systems are obviously utilized in set-shifting.

Assuming that one's set-shifting, attention, and memory systems are intact, skill in the planning ability is developed through experience and through the capacity for reflective judgment. This ability begins with the capacity to realize goals in the near future within relatively stable contexts. If a child's goal is to acquire a camera, and her plan for obtaining it is to ask her parents, such a strategy occurs within a relatively stable context depending on few variables and contingencies. But as one matures, one's goals require more complex strategies because they depend on a greater number of contingencies. For example, one's goal may be to purchase a home by the time one is 30 years old. Such a goal depends on number of variables and uncertainties (one's job prospects, housing prices, the credit market, etc.). When one's plan is sensitive to the uncertainties and possible obstacles that may thwart it, one attains a more sophisticated planning ability. This more mature planning ability grows out of experience and the ability to reflect on this experience so as to make it useful to one's pursuit of future goals.

As mentioned above, in addition to the intellectual and planning abilities of the executive function, goal pursuit depends upon competencies that enable one to implement the plan and see it to fruition. These competencies fall under the two other components of the executive function: the navigational and the motivational. Let's now turn to the navigational component.

The Navigational Component of Executive Function

The navigational component of the executive function is fundamental to goal attainment, especially where the goals are long term ones.

Essentially, the navigational component contains the competencies that one needs to *execute* one's plan. This is no easy task and requires a very important ability known as *self-regulation.*

Self-regulation

Self-regulation is at the heart of, and is absolutely essential to, any type of goal attainment. It allows one's body to "stay on target" despite the ever changing circumstances of human affairs. As described by scientists, self-regulation consists in the ability to "continuously monitor progress toward a goal, checking outcomes, and redirecting unsuccessful efforts."[10] Self-regulation not only includes the set-shifting ability, but also includes the ability to restrain oneself from acting on impulses that jeopardize the attainment of one's goal, an ability known as *inhibition.*[11] This inhibitory ability is developed from a more fundamental capacity known as "private speech."

Private speech

Private speech refers to the use of speech to communicate to oneself rather than to others. Regarded as a function of language acquisition that begins to first develop in the preschool years, the "seed" from which private speech grows is thought to be social interaction with other individuals. Language begins as a tool—rich with rules and symbols—which is used to communicate with others. As the child matures, however, he or she begins to use language to talk to him or herself. The first displays of this "self-talk" are external—we have all witnessed that frequent act of "talking out loud" that pre-school children exhibit. This speech display is called "external self-talk" and it is most ubiquitous during "task performance"—when the child is engaged in a goal-oriented action such as solving a puzzle. It is thought that what the child is doing here is providing self-direction just as an adult would, if an adult was helping the child to solve the puzzle.[12] Studies show that as the difficulty of the task increases, so does the frequency of self-talk, and this finding has led to the conclusion that children use self-talk to overcome obstacles.[13] Just as a body builder may groan to help him lift a heavy barbell, children will talk to themselves when a task requires extra concentration or mindfulness.[14]

Eventually, task-oriented external self-talk "goes underground," becoming less overt as the child matures. The external self-talk becomes internalized and private, thereby developing into *private speech.* Through private speech, the child is able to use the symbolic tools of her language to set and plan her *own* goals and ends. At this stage of development, the child uses language internally and privately. If a child's goal is "read Aesop's Fables," and if, on her own, she engages in the actions necessary to achieve this end, she begins to engage in a more sophisticated use of private speech. Private speech is important to the executive function not only because it provides the tools and symbols that are utilized by the planning component, but also because it is involved in the development of self-restraint that underlies the inhibitory element of self-regulation. To see how, we need to discuss the development of private speech. For private speech is *itself* a developed capacity.

The development of private speech

A well-developed private speech capacity emerges from specific environments. Research demonstrates that children who grow up in relatively silent, taciturn households have an impaired private speech ability.[15] Thus, environments with robust discourse nurture the capacity for private speech. Without sufficient exposure to language, a child's internal "tool-chest" is impoverished. In addition to environments rich with discourse, environments that provide a child with ample opportunity to engage in "free" or unstructured play are also crucial to the development of private speech.[16] Free play is interestingly linked to the development of self-restraint and inhibition.

Free play typically involves the rules and norms of the community, and so a children's game of "playing dentist" incorporates the norms governing this role. During free play, the child learns to restrain her impulses for the sake of conforming to the norms that govern the play.[17] This activity is thought to have a developmental function as, through it, the child experiments and practices with the voluntary exercise of will power and impulse control. In fact, in preschool children, will power and self-restraint are at their highest during make-believe play.[18] Some researchers have linked the rise in ADHD to the decline in free play.[19] Because free play allows the child to practice the restraint of her impulses, it plays an important role in the development of self-regulation.[20]

To sum up our discussion thus far, we have seen that environments rich with discourse and opportunities to engage in free play nurture the development of private speech. Free play, moreover, facilitates the development of the inhibitory competencies involved in self-regulation by allowing the child to experiment and practice impulse control in the context of play. But fully developed self-regulation implies that the individual can restrain and inhibit one's impulses *outside* of the context of play. And as we will see, private speech aids in this further development and extension of self-regulation.

The steering function of private speech

Laura Berk, a leading researcher on private speech, observes that private speech is used throughout one's life, especially during challenging moments which require extra energy and/or attention.[21] At such moments, the individual uses private speech outside of a play context and applies it to adult life. I call this mature use of private speech the *steering function* of private speech. This sophisticated ability consists in the individual's conscious and deliberate use of private speech in order to inhibit and restrain impulses in non-play contexts.

The steering function of private speech is absolutely essential for self-concordance as realizing one's chosen end requires one to inhibit and restrain one's impulses that thwart one from this end. For example, if one's self-concordant end is "being a philosophy professor," then one must restrain oneself from performing those actions that undermine the attainment of this goal. Thus, as one pursues a graduate degree in philosophy, one must restrain oneself from, for example, going to the bar instead of writing the seminar paper that is due the next day. Self-regulation, through the steering function of private speech, is the capacity in virtue of which one exercises this restraint. Upon being tempted by the prospect of going to the bar, one exercises the steering function of private speech when one says "no, you need to write the seminar paper" and *listens*.

The steering function of private speech enables one to overcome something called "stimulus control."[22] Stimulus control is *bad*. It amounts to a *failure* in self-regulation. When one is a victim to stimulus control, one is "thrown off course" because one fails to exert sufficient will power against a distracting, discouraging, and/or tempting stimulus. The counterweight against stimulus control is the steering function of private speech. But importantly, if used regularly, the steering function of private speech can help one to develop a more habitual, regular self-regulatory ability. Each usage of the steering function of private speech has developmental consequences. The more one engages in it, the more regular and habitual it becomes. And the more regular and habitual it becomes, the greater is one's capacity for self-regulation. Put another way, the repeated use of the steering function of private speech *awakens* one's capacity for habitual self-regulation. As some modern researchers describe it, self-regulation is a "muscle" that can be developed with training. And it is by repeatedly using the steering function of private speech by which one develops the "self-regulation muscle."[23]

Aristotle on private speech

The steering function of private speech should remind you of our discussion of affective virtue from Chapter 3. Recall that the cultivation of affective virtue involves both a private and a public component. The private component of this training consists in repeatedly listening to a rule or principle that is grasped by practical reason. The individual who chooses to develop courage repeatedly says to himself, "I desire to flee, yet the rule says that I must stand my ground. I will listen to the rule." And the individual who chooses to develop temperance repeatedly says to himself, "I am tempted and my desire is luring me to the forbidden act, yet the rule says that I must say no. I will listen to the rule." The development of affective virtue requires repeated acts of will power.

The steering function of private speech underlies, and is at the heart of, the private component of affective training. For it is by repeatedly listening to the rule or rational principle, and hence using private speech to direct the body when desire and impulse pull the other way, whereby one awakens and develops the virtues. The training and cultivation of affect involves and depends upon repeatedly controlling one's impulses by listening to a rule. And by repeatedly listening to the rule, one eventually *habitually* listens to the rule. Indeed, Aristotle tells us that the virtuous man is one whose actions regularly conform to the correct rule or rational principle:

> Now, that we must act according to the right rule is a common principle and must be assumed. (*NE* 1103b)
>
> [T]he brave man feels and acts according to the merits of the case and in whatever way the rule directs. (*NE* 1115b)
>
> [B]oth the continent man and the temperate man are such as to do nothing contrary to the rule for the sake of the bodily pleasures. (*NE* 1151b–1152a)
>
> [T]he good-tempered man tends to be unperturbed and not to be led by passion, but to be angry in the manner, at the things, and for the length of time, that the rule dictates. (*NE* 1126b)
>
> [Affective virtue], then, is a state of character concerned with choice, lying in a mean, i.e. the mean relative to us, this being determined by a rational principle, and by that principle by which the man of practical wisdom would determine it. (*NE* 1106b–1107a).

In these passages, Aristotle alludes to the habitual operation of private speech that is characteristic of virtuous individuals. "Obeying the rule" is an act of private speech, and the mark of a virtuous person is to *habitually* obey the correct rule. To develop this habitual obedience, as we know, requires one to repeatedly use the steering function of private speech. Similarly, to acquire habitual *self-regulation*, one also repeatedly uses the steering function of private speech. Aristotle's account of the development of affective virtue mirrors the development of self-regulation.

Self-regulation can be so strengthened and developed that it becomes *masterful*. The development of masterful self-regulation involves further extensions of the steering function of private speech. One important extension involves using private speech on specific types of habits. Let us begin to discuss masterful self-regulation by focusing on the connection between self-regulation and habit.

Self-regulation and habit

If you recall from Chapter 3, Aristotle believes that training is necessary to correct man's default affective state. I said there that man's default affective state was entirely natural given the fact that human beings begin their first few years acting at the beck and call of impulse. Recalling William James' remarks about the plasticity of the nervous system, and Aristotle's similar claim that one's repeated actions leave their mark on one's character (*NE* 1114a), it certainly makes sense why obedience to reason must be trained. Because virtually all of one's actions are impulse driven for the first few years of life, a corresponding character and motivational system results. Because this character and motivational system does not conduce to flourishing, a correction is needed. The steering function of private speech is essential to this correction. Through private speech, the individual overcomes and changes habitual behavior that does not conduce to flourishing.

The ubiquity of automaticity

Aristotle's insight into the importance of habits has been borne out by research on habitual or automatic behaviors. Habitual behaviors are automatic responses to certain situational factors in one's environment.[24] In contrast to behavior that is deliberately intended, habitual behavior is learned behavior that is automatically triggered by certain environmental cues.[25]

Although it may seem that most human behavior is the result of deliberate intent and volition, current research demonstrates that a large percentage of our behavior occurs automatically, and without deliberate intent. On some estimates, approximately 50% of our behaviors are automatic and habitual.[26] Others estimate this number at 95%.[27] William James writes "ninety-nine hundredths, or, possibly nine-hundred and ninety-nine thousands of our activity is purely automatic and habitual."[28]

To appreciate the ubiquity of habitual actions, reflect, for a moment, on your experience of reading. Doubtless you have had the experience of "daydreaming" while reading. Suppose you have said to yourself, "I am going to read *Anna Karenina.*" You find a quiet spot and begin to read. What typically happens is that as you read, certain words or sequences thereof trigger your thoughts or memories, which themselves trigger further thoughts or memories. Before you know it, you have spent the last 10 minutes away from the words on the page, daydreaming about your summer vacation in Charleston. It all happens automatically, somewhat innocently and without deliberate intent. This is but one illustration of how words and sentences can trigger—without will, volition, or deliberate intent—memories that divert one from one's goal of reading *Anna Karenina.*

For another example, consider the phenomenon of not remembering where you parked your car when you get to the parking lot to find it. This happens because when you parked your car your attention was elsewhere because it *could* be. You are so adept at parking the car and walking from the lot to your destination that the "script" to get to the building occurs quite effortlessly, without concentration or will power. Upon parking your car, you may have been singing along with the radio, or talking on your cell, totally absorbed in some other activity. The problem is, in failing to be mindful when you parked the car, you frustrate your goal of finding the car when you return to the lot.

Now, not all habitual behaviors are bad. Aristotle, recall, claims that habitual action plays an essential role in affective virtue. And William James exhorts us to develop efficient habitual routines so that we can devote our attention to the things that matter. He writes "the more of the details of our daily life we can hand over to the effortless custody of automatism, the more our higher powers of mind will be set free for their own proper work."[29] Habitual, automatic behavior enables us to be more efficient. But the trick is to acquire the habits that facilitate the attainment of our goals, and to eliminate those habits that interfere and thwart us from our goals. When one can do this, one has developed *masterful* self-regulation.

Masterful self-regulation: mindfulness of habit triggers

The ability to overcome situational factors that trigger responses which interfere with one's goal first and foremost requires mindfulness and awareness. To overcome stimulus control, one must first be mindful of not only the offending habitual behavior, but also of its trigger. In my own experience, I noticed that, although I sat down at my desk with the deliberate goal of working, my goal was often thwarted when I decided to check my email. Eventually, I recognized that this seemingly innocent and simple act could swallow up a significant portion of time. How? Because checking my email triggered a whole suite of habitual "web-surfing" behaviors (check cnn.com, click on links, check Twitter, click on links etc., look at the status of my orders on Amazon.com, click on links, etc.), and at that point, I was off and running sometimes losing hours of work. Sometimes, it takes a while to recognize that a certain event is triggering a sequence of unwanted behaviors, but this is the first and crucial step to overcoming stimulus control. Awareness and mindfulness of the trigger is key, and is the first step in developing a masterful self-regulatory ability.

Masterful self-regulation: classification of habit triggers

Once one is aware that some factor is a trigger of damaging or threatening behavior, a further determination must be made about whether the trigger is *controllable* or *uncontrollable.* Doing so determines which self-regulatory skill should be employed. If the situational factors are controllable, then they can be removed from one's environment so that the automatic habits are less likely to be triggered. If, on the other hand, situational factors are *un*controllable, and hence removing them from the environment is not an option, then a different self-regulatory skill is needed.

For example, suppose Amy's goal is weight loss and she is accustomed to eating cookies as a late night snack. In fact, when she begins, she often eats 20–25 cookies at a time. For her, eating just one cookie is a trigger for eating many, many more. But because Amy controls what foods are available as a late night snack, she can control this situational factor. So, after she becomes mindful and aware that the presence of cookies threatens her goal of weight loss (hopefully, this will dawn on her rather quickly), the self-regulatory competency that she employs is *engineering her environment* so that it does not trigger her late night habit. Amy may throw out the cookies, or bring them to work to give to her co-workers, and she may refuse to purchase them in the future in order to steer herself towards her goal. In my own case, since I can control when I check email, I have enacted a policy of turning off my wireless connection when I work. In so doing, I have "removed" the stimulus (being able to check email) that triggers automatic behaviors that interfere with my goal. Research demonstrates that removing the triggering stimulus from the environment undermines the occurrence of the corresponding automatic behavior.[30]

Sometimes, however, we cannot remove the triggering stimulus from the environment. Suppose that Amy works for a company which provides pastries and coffee to its employees every morning. For the past several years, Amy has availed herself of this perk. Even worse, the pastries are set out in a mouth-watering display that she must daily pass to get to her office. Further suppose that the situational factors are *un*controllable and hence that asking her company to stop the service or move the display is not an option. In this situation, the self-regulatory competency needed to overcome stimulus control involves using the steering function of private speech. This is a very sophisticated use of the steering function of private speech for one uses it to *steer attention.*

Masterful self-regulation: steering attention

Research on will power or "delay of gratification" demonstrates that acts of will power involve a strategic use of one's attention.[31] To understand how this works, it helps to first think of the Rubin vase/profile illusion. Why? This illusion (Figure 5.1 pictured below) demonstrates that our attention determines what we see. If our attention focuses on the black portions of the image, we perceive two profiles. But if

FIGURE 5.1 The Rubin vase/profile illusion

our attention focuses on the white portion of the image, we perceive a vase. By shifting our attention we can see things differently. This idea is at the heart of the strategy for overcoming uncontrollable factors that trigger damaging habits.

Researchers have divided the situational factors that trigger interfering habits into two "attentional fields." Specifically, attention can either be focused on the "hot" qualities that comprise the "action arousing" field, or it can focus on the "cool" qualities that comprise the "arousal dampening" field. Think of the fields simply as different bundles of qualities. "Hot" action arousing qualities of a piece of cheesecake, for instance, would be its creaminess and its richness. By focusing on these qualities, we exacerbate the temptation, thereby exciting arousal. But there are qualities of the cheesecake that, if focused on, dampen arousal. For example, we can focus on the shape of the cheesecake, or its color. Instead of seeing it as a mouth-watering luscious, creamy dessert, we can see it as a yellow circular mass. In so doing, we shift our attention to the "cool" qualities of the cheesecake, thereby dampening its ability to tempt us.

Using self-distraction or focusing on "cool" non-action arousing attributes of the stimulus, are two methods which enable one to demonstrate will power when confronted with a potentially controlling stimulus.[32] In one experiment, children who thought of marshmallows as "puffy round clouds" or who distracted themselves from the treats, were better able to refrain from eating them than were children who thought the marshmallows as "sweet and chewy."[33] Thus self-distraction and shifting attention to the "arousal dampening" qualities, are the mechanisms that enable one to overcome factors within one's environment that trigger interfering habits. Both strategies make use of the steering function of private speech. To overcome her desire for a morning pastry, therefore, Amy may look away as she walks past the table and say to herself, "I need to distract myself and so let me recite the Greek alphabet," or she may look at the pastry table in terms of its abstract, geometrical qualities ("there are 20 light brown disks on that table"). In both cases, private speech is used to steer Amy's attention to aspects of the situation which serve to dampen her arousal.

Masterful self-regulation: steering attention inwards

It should also be pointed out that factors that trigger interfering habits do not have to be external. Indeed, they can be *internal.* There may be habits of thought or interpretation that interfere with our goals. And here, too, the steering function of private speech can be used to overcome these internal interferences. Again, the first step in being able to use the steering function of private speech involves mindfulness and awareness of the thought patterns and interpretations that interfere with goal attainment. One needs to think about one's own thinking and one's own interpretations. Meta-cognition, thinking about one's own thinking, is an important use of private speech that enables an individual to avoid her own cognitions that undermine her ability to set and/or realize goals.

A contemporary illustration of how private speech can be used to overcome "interfering cognitions" can be found in the story of John Nash, the brilliant mathematician who is the subject of the book and the movie *A Beautiful Mind.* Nash, who suffered from hallucinations and delusions consistent with schizophrenia speaks of a "constant struggle" to rid himself of the delusional habits of which he was a victim. These internal interferences derailed Nash from both personal and professional goals. But through the steering function of private speech, Nash was able to overcome these impairments. The following description of Nash's technique alludes to his use of this sophisticated self-regulatory skill:

> Nash has described his illness not in terms of highs and lows, bouts of mania followed by disabling depression, bur rather in terms of a persistent dreamlike state and bizarre beliefs . . . He has spoken of being preoccupied by delusions, of being unable to work, and of withdrawing from the people around him. Mostly he has defined it as an inability to reason. Indeed, he has told . . . others that he is still plagued by paranoid thoughts, even voices, although, in comparison to the past, the noise level has been turned way down. Nash has compared rationality to dieting, implying a constant struggle. It is a matter of policing one's thoughts, he has said, trying to recognize paranoid ideas and rejecting them, just the way somebody who wants to lose weight has to decide consciously to avoid fats or sweets.[34]

Nash's attempt to "recognize and then reject" paranoid ideas certainly involved the use of the steering component of private speech. Note that his strategy involved both "recognition" and "rejection." Nash first became mindful and aware of the offending ideas, and then dismissed them. To dismiss them, he focused his attention on certain qualities of these thought patterns in an act of meta-cognition. Upon observing these harmful thought patterns, he probably said something like "that is an unhealthy thought, I need to focus my attention elsewhere. Here, let me focus on proving this theorem." By being mindful of his own cognitions, he was able to appreciate how they affected his well-being. This the first crucial step, *being aware that your cognitions can derail you from your goals.* When Nash understood that some of his cognitions were hindering him from his goals, he employed the steering function of private speech to direct his attention away from them. Using private speech to steer attention away from the internal cognitions that derail is certainly a masterful use of self-regulation.

Although Nash's mental illness is an extreme version of an internal habit that can compromise and interfere with one's goals, the fact is that most individuals have cognitions—thoughts and interpretations—that can thwart them from their goals. These cognitions can be distracting and/or debilitating. The Stoic school of philosophy recognized this and exhorted individuals to tend to their judgments, for it is judgments—one's interpretation of things—and not the things themselves that gives rise to negative emotions and distress.[35] For example, Cicero mentions focusing attention on the sentence, "you are not the only one," or on images of how sensible men have dealt with death and loss, to help mollify one's grief.[36] Again, the idea is to shift attention to an idea or image that helps to dampen the harmful effects of certain cognitions.

One debilitating habitual cognition concerns one's *causal interpretation* of bad events such as a break-up or other misfortune. Studies show that one's "explanatory style," that is, the habitual manner by which one interprets the cause of a bad event, is a risk factor for depression.[37] Essentially, if one's interpretive habit of a bad event focuses on internal, stable, and uncontrollable factors, one is more prone to depression.[38] But such debilitating interpretive habits can be overcome and one can learn to interpret bad events by focusing on external, changeable, controllable factors. Instead of interpreting a break-up by saying something like "she left me because I am unlovable and a loser," one can instead focus on the external, changeable, and controllable factors of the situation and say something like "she was not ready for a serious commitment. I will be more careful next time." Changing one's explanatory style involves the steering function of private speech. One first learns to *recognize* one's explanatory style, and then one uses private speech to shift attention to qualities of the bad event that are external, changeable, and controllable.[39] Doing so helps the individual to steer herself towards her self-concordant goals, instead of focusing on painful cognitions that undermine one's resolve to realize them.

Masterful self-regulation: forming implementation intentions

Masterful self-regulation also includes the ability to form *implementation intentions.* Coined by Peter Gollwitzer, the term "implementation intention" is distinguished from a *goal intention.* A goal intention is *broad* such as "get an A" or "Lose 10 pounds." Unlike goal intentions, implementation intentions are narrow and highly specific "mini-goals" that "specify the when, where, and how of responses leading to goal attainment."[40] Recalling our discussion about the craftsman's motivational system from Chapter 2, we noted that Aristotle distinguishes *desire for the end* (wish or calling) from *desire for the means* (choice) (see pages 26–32). In order to make a silver pen, Namiki's calling, he has to desire this calling *and* the specific means by which it is achieved (going to the market to get silver, etc.). Choice, recall, refers to the desire for the specific means. Implementation intentions are akin to choice—they refer to the intentions to do very specific things, in very specific circumstances, so as to ensure that one's goal intention is achieved. Research shows that goal intentions that are supplemented by implementation intentions are more likely to be realized.[41]

Understood within the context of self-regulation, the ability to form implementation intentions is a powerful skill that assists one in overcoming and undoing interfering habit triggers. For example, suppose Jason's goal is "*get straight A's this semester,*" and further suppose that his roommates often interfere with his studying by tempting him with the latest video games. Jason then needs to form a specific strategy—a specific "mini-goal"—given that this circumstance threatens to derail him from his goal of getting straight A's. If he has masterful self-regulation, Jason will form an implementation intention such as "*if my roommates begin to distract me from my studies by tempting me to play video games, I will get my earplugs and put on my earphones.*" Note how this intention specifies a mini-goal ("get my earplugs and put on my earphones") given a specific situation that threatens to derail Jason from his main goal. Indeed, the mini-goal is formed *in the service of* Jason's main goal of getting straight A's. To use another example, if Amy's goal is "*lose 10 pounds,*" and if her repeated visits to the donut display at her workplace have derailed her from her goal, she might form the following implementation intention, "*if I am tempted at work by the donut display, then I will take out a stick of gum.*" The following table breaks down both Jason's and Amy's implementation intentions into the triggering circumstance and the specific mini-goal designed to overcome that circumstance:

Main goal	**Threatening situational factor**	**Mini-goal designed to overcome situational factor so that main goal is achieved**	**Implementation intention**
Jason: get straight A's this semester	Roommates distracting me and tempting me with video games	Get my earplugs and put on my earphones	*If my roommates begin to distract me from my studies by tempting me to play video games, I will get my earplugs and put on my earphones*
Amy: lose 10 pounds	Donut display at work	Chew gum and walk away	*If I am tempted by the donut display at work, I will take out a stick of gum and walk away*

The use of implementation intentions is a yet another component of masterful self-regulation. Indeed, one who has developed masterful self-regulation can use implementation intentions to aid and assist the steering function of private speech to overcome habit triggers. Note that when one uses implementation intentions in this way, one *recognizes* one's own weakness in the face of specific situational factors. It is because Jason understands that video games threaten his goal of straight A's which enables him to develop a strategy for overcoming this obstacle. And it is because Amy recognizes that the donut display at work threatens her goal of weight loss that enables her to tackle this problem with an implementation intention. Thus, using implementation intentions to assist private speech implies that one has *experience with* and is *mindful of* one's weaknesses in the face of certain environmental triggers. The mindfulness and awareness of one's weaknesses, the environmental cues that trigger them, *and* the formulation of a specific strategic goal for overcoming these weaknesses, enables one to more effectively realize one's ends.

Now, our discussion of implementation intentions has been focusing on *external* interfering habit triggers: distracting roommates, video games, mouth-watering donut displays, and so on. But as you now know, interfering habit triggers can also be *internal.* And implementation intentions can be used to overcome these internal obstacles as well. As discussed earlier, one can use private speech to direct attention away from distressing cognitions and damaging habits of interpretation. In an act of meta-cognition, one can use private speech to "steer attention inwards" and hence to thoughts, ideas, and images that soothe, energize, and inspire. And using implementation intentions can help one to achieve such "internal mastery." Suppose Liz becomes mindful of her faulty "explanatory style" in which she habitually interprets bad events as stemming from unchangeable and uncontrollable factors. For instance, when she gets a bad grade she thinks "I am stupid and worthless." Further suppose that she desires to sculpt a healthier explanatory style. Using implementation intentions can help Liz to achieve this goal. In this case, the "situational factor" to overcome is *internal* and is Liz's own pessimistic explanatory style. An implementation intention designed to overcome this "internal hurdle" may be: *if I see myself thinking "I am stupid and worthless" when I get a bad grade, then I will make a mental list of things that I can do to change my study strategy so that I get a better grade next time."* So:

(Internal) main goal	**Threatening (internal) situational factor**	**Mini-goal designed to overcome "internal hurdle" so that main goal is achieved**	**Implementation intention**
Liz: Develop a more healthy and optimistic explanatory style	My explanatory style in which I habitually interpret bad grades as a sign that I am stupid and worthless	Make a mental list of things that I can do to change my study strategy so that I get a better grade next time	*If I see myself thinking "I am stupid and worthless" when I get a bad grade, then I will make a mental list of things that I can do to change my study strategy so that I get a better grade next time*

The use of implementation intentions on threatening internal habit triggers also presupposes mindfulness and the recognition that one's own cognitions are interfering with one's ends. Recalling our discussion about how shifting one's attention to "cool, non-action arousing attributes" enables one to overcome a triggering stimulus, we can see that one who uses this technique consciously and deliberately uses a specific implementation intention in which the mini-goal is an *internal* one, namely "*shifting attention to the cool attributes.*" The important point is that neither the interfering triggers nor the mini-goals designed to deal with them need be "external." Both trigger and mini-goal can be internal. Using implementation intentions to overcome threatening stimuli—whether external or internal—is a form of self-mastery and is a manifestation of masterful self-regulation.

To summarize the various competencies underlying masterful self-regulation, we said that the first competency involved in avoiding interfering habitual behaviors is mindfulness and awareness of the trigger. One then needs to make a further determination as to whether the trigger is external (e.g., in one's environment) or internal (e.g., a cognition or interpretation), and whether it is controllable (e.g., cookies in one's pantry) or uncontrollable (e.g., free donuts at work). The classification of the habit trigger determines which self-regulatory skill to use, and helps one to form the requisite implementation intention. The following table summarizes some of the different skills that can be utilized.

Masterful self-regulation: overcoming the interfering habit trigger

	Controllable	**Uncontrollable**
External	Engineer environment to remove the trigger (e.g., Amy removes the cookies from her pantry).	Employ the steering component of private speech and form an implementation intention to self-distract or shift attention to arousal dampening qualities of stimulus (e.g., Amy says to herself, "*if I am tempted by the pastry display at work, I will focus my attention on the arousal dampening attributes and look at the pastries as beige disks*").
Internal (distressing cognitions, unhealthy explanatory style)	Employ the steering component of private speech and form an implementation intention to shift attention to the controllable factors of the situation or to cognitions that conduce to more positive affect (e.g., Liz says to herself, "*if I see myself thinking "I am stupid" after I get a bad grade, then make a mental list of different study strategies that I can try out next time*").	N/A. "To choose one's attitude in any given set of circumstances is the last of the human freedoms." (Victor Frankl, *Man's Search for Meaning*)."[42]

Masterful self-regulation and self-concordance

The fact that habitual, automatic actions are a necessary component of the human condition has obvious relevance to self-concordance. We may consciously and deliberately set a self-concordant goal for ourselves (e.g., teach biology for a living) but nevertheless get derailed from our goal because the external features of our environment and/or our internal cognitions trigger automatic interfering behaviors. Unless the individual can access "specific self-regulatory competencies,"[43] numerous factors can trigger habits that undermine the attainment of our goals, despite our sincerity and resolve in setting them.

Many self-concordant goals—especially those that involve professional aspirations—are long-term goals. As such, their realization requires the prior realization of a number of "mini-goals," or, to use Aristotle's expression, "hypothetical necessities." For example, if one's self-concordant goal is to become a philosophy professor, then one must first graduate from college, and then complete graduate school. But the realization of each of these goals requires one to first realize a huge number of mini-goals or hypothetical necessities (applying to college, enrolling in courses that fulfill one's degree, taking exams, writing papers, applying to graduate school, enrolling in courses that fulfill one's degree, etc.). But as one begins to undertake the necessary steps needed to realize one's self-concordant goal, one may feel far removed from the big picture. The self-concordant goal is in the distant future, and our limited attention is absorbed in the "mini-goals" which we must first realize to get there. And so one may become absorbed in the pressure of one's required coursework and forget why one is doing what one is doing. When this happens, we are especially vulnerable to interfering habits.

Even worse, many self-concordant goals require college degrees, and college degrees are heavily laden with externally imposed goals, some of which have no obvious relevance to one's self-concordant goal. Moreover, "bad events" such as doing poorly on an assignment can trigger interfering and derailing cognitions ("I am a loser, I do not belong here"). In such an environment, the danger of interfering habits looms large. Without masterful self-regulation, one can easily veer off course.

But even if our self-concordant goals do not require college, masterful self-regulation is clearly necessary for true self-concordance. Automatic behaviors are a feature of a human condition. So are bad events. And unless one was raised by the Buddha himself, one has automatic behaviors (triggered by both external and internal factors) that interfere and derail. In order to construct a life project that is "fully packed" with self-concordant activities, one needs masterful self-regulation. Masterful self-regulation implies the ability to be aware of one's interfering habits, *and* the ability to classify and categorize their triggers. Put simply, masterful self-regulation implies "self-knowledge" of one's own habits and their corresponding triggers—whether external or internal. But this is not all. Masterful self-regulation also implies the ability to *use* this self-knowledge to formulate the appropriate implementation intentions through which one resculpts and reshapes one's habits. Masterful self-regulation implies a suite of abilities that enable one to develop habits that facilitate rather than hinder the realization of one's ends. Clearly, masterful self-regulation is necessary to construct a life project that has maximal self-concordance.

Now, even though one possesses masterful self-regulation, this is not sufficient for self-concordance. One also has to commence the actions that are necessary for the realization of one's self-concordant end. One has to *begin* and, as Nike says, "just do it." This ability refers to third and final component of the executive function, the *motivational* component.

The Motivational Component of Executive Function

The motivational component of the executive function concerns those competencies that motivate and engage the body. I call the engagement of the body *somatic engagement*. This may seem like an unimportant aspect of executive function, but it is absolutely essential for goal attainment. Our body does not obey our commands so easily. Procrastination abounds, setbacks discourage, and there is apathy.

While there are a number of things that can contribute to somatic engagement such as the intrinsic desirability of the goal, intrinsic desirability may not be enough to actually motivate one to pursue it. One also needs an underlying motivational belief that the goal is obtainable. Research within goal attainment theory demonstrates that there are a variety of factors that contribute to this motivational belief.

Framing

For example, framing the goal matters: if the goal is understood not as something insurmountable, but is broken down into smaller, more manageable steps, it is more likely to be achieved.[44] In order to be effective, the motivational belief must "frame" the goal in a manner that reduces it to its simpler components. In the film *The Legend of Bagger Vance,* Junnah, a rookie golf player, is losing a tournament to two professionals, Bobby Jones and Walter Hagen. Discouraged, a child who is rooting for Junnah (who is 12 strokes behind) explains to him that beating the champions is simple: since there are 54 holes left in the game, all Junnah has to do is pick up one stroke every four holes and he wins. Framing a seemingly impossible task ("beat the two champions while 12 strokes behind") in terms of the more manageable "pick up one stroke every four holes" exemplifies the framing that conduces to motivation. The marathoner's adage "It's not 26 miles, it's one mile 26 times" makes the same point.[45] If one's task is not "insurmountable" but is just plain old intimidating (e.g., studying for an exam), framing helps to induce motivation. Instead of framing the goal as "study for exam" one can frame the task as a set of manageable, doable goals such as "study for 90 minutes." This goal has a clear endpoint and is much less intimidating that "study for the exam." In addition to inducing motivation, framing helps to improve performance when the goal is framed as something very specific. The more specific a goal is, the more focused one's attention is.[46] The goal "study for 90 minutes" while more specific than "study for the exam," could be made *even more* specific

not concerned with habit triggers here, but with the ability to move forward despite great adversity. An individual who seeks a professional degree, for example, may encounter severe set backs such as the death of a spouse or a child. If she is resilient, she will press on to realize her ends, despite such adversity. Recent research on resilience suggests that resilience is quite "ordinary," and results from "ordinary human adaptive processes."[52] In other words, all human beings have the potential for resilience. The trick is to create ecologies that nurture, promote, and actualize these natural adaptive processes, and to eliminate those ecologies that thwart them.

Masterful motivation: using private speech to motivate

One can use private speech to call on or direct attention to specific beliefs or images for the purpose of inducing motivation. For example, if one is mindful of the importance of framing, then when one encounters a seemingly insurmountable goal, one can say to oneself "I need to break this down into smaller components" thereby commencing the internal search for the right frame.

The act of calling on specific inner resources in order to induce motivation is yet another extension of the steering function of private speech. But in this case, one uses private speech not to restrain but to push, to motivate. Using private speech in this way is a special use of the "steering attention inward" component of private speech that was discussed in connection with masterful self-regulation. But again, in this case, private speech is used to steer attention towards certain cognitions for the purpose of inducing motivation. This may include framing, it may include recalling a story about someone overcoming adversity so as to bolster hope and self-efficacy, it may include the deliberate decision to listen to specific music, to watch a certain film, to dance, and so on. But whatever the technique used, one uses the steering function of private speech to stimulate motivation when needed. I refer to this use of private speech as *masterful motivation*. Like masterful self-regulation, masterful motivation makes use of implementation intentions. Note that "search for the right frame" or "think about the determination of Susan B. Anthony" could be the goal of an *implementation intention* that an individual crafts in order to induce motivation in the face of circumstances that threaten it. When one knows which specific strategies to employ for the sake of inducing motivation, one can form a specific implementation intention aimed at producing the motivation and energy necessary to realize one's ends. Obviously, the specific strategies that work will vary from person to person: for some thinking of Susan B. Anthony will do nothing, for others, it will ignite the spark needed for motivation. The important thing is that one discovers which technique works for inducing motivation when needed. The table below summarizes how one can use implementation intentions to develop masterful motivation.

(Internal) main goal	**Threatening (internal) situational factor**	**Mini-goal designed to induce motivation**	**Implementation intention**
Induce Motivation to develop masterful study strategies	Feeling depressed and unmotivated after failing an exam.	Listen to the song "Free Will" by Rush, watch the film *Chariots of Fire*, and/or think of Susan B. Anthony and read some of her letters to Elizabeth Cady Stanton.	*If I realize that I need to induce motivation to continue to develop masterful study strategies after failing an exam, then I will listen to the song "Free Will" by Rush, watch the film* Chariots of Fire, *and/or think of Susan B. Anthony and read some of her letters to Elizabeth Cady Stanton.*

Summarizing our discussion thus far, we have been discussing some of the various competencies and strengths that are necessary for an excellence in *goal attainment.* Practical reason, as we have repeatedly said, is the rational capacity responsible for goal-oriented activity. And as we are learning, *excellent* practical reason requires a number of strengths and competencies that are made apparent when we consider the various components underlying the executive function. Some of these strengths are analytic strengths such as the ability to engage in strategic planning and hence to think backwards from end to means, thereby formulating an action plan that is sensitive to the variability and contingency that is characteristic of practical affairs. Other strengths concern the self-awareness and mindfulness that are necessary to overcome automatic behaviors that thwart one from her goals. There are "navigational" and "steering" competencies that, when perfected, imply a nimbleness in the steering component of private speech. Still other strengths concern affect and motivation: the ability to frame the goal in sufficiently proximal terms; hope and self-efficacy; resilience; and the masterful ability to use private speech to induce motivation. As should be apparent, if cultivated and developed, private speech and implementation intentions are central to excellence in goal attainment. Together, they function as an "inertial guidance system,"[53] restraining us and pushing us as is needed to realize our goals. The following table provides a handy summary of our discussion of the capacities required for excellence in the ability to set and realize goal, i.e., teleological excellence.

Capacities necessary for teleological excellence

Analytic Competencies	Calculating/strategic planning Mindfulness of both external and internal habit triggers; Classification of habit triggers as controllable/uncontrollable.
Navigational Competencies	Using the steering function of private speech assisted by implementation intentions to: a) shift attention to arousal dampening qualities; and b) cultivate an explanatory style such that the causes of "bad events" are attributed to factors that are controllable and changeable; and c) shift attention to cognitions that induce positive affect.
Motivational Competencies	Using the steering function of private speech assisted by implementation intentions to induce motivation that leads to the implementation of the strategic plan via physical action of the body; Framing, hope & self-efficacy, resilience

To engage in self-concordant activity requires, at the very least, the analytic, navigational, and motivational competencies listed above. Self-concordance requires that one set and devise the strategic plan by which the self-concordant goal will be realized. Because many self-concordant goals are long-term professional pursuits, one needs masterful self-regulatory competencies that enable one to stay on course and overcome the control of factors that trigger interfering behaviors. One will need a reservoir of techniques and skills to help one inhibit and exhort, restrain and push, when needed.

The analytic, navigational, and motivational strengths listed above can be thought of as *virtues*. For these strengths underlie an excellence in goal attainment. If flourishing is a specific type of excellence in goal attainment, then it will require the development of the analytic, navigational, and motivational virtues mentioned above. And since self-concordance is a type of excellence in goal attainment, then *it* requires the development of these virtues.

One crucial ingredient that has been left out of the list of preconditions of self-concordance is, of course, the ends or goals that provide direction for the above competencies. As mentioned earlier, these ends or goals will vary from person to person, and are specific to an individual's enduring values and interests. There is some skill required in understanding what these enduring values are for oneself. Indeed, these values are not transparent and require asking oneself the right *questions*. Some questions that have been used with great success include "how do you want to be remembered?" and "if you could somehow watch your own funeral, what would gladden you to hear?"[54] So in addition to the aforementioned skills, self-concordance also requires the ability to know one's enduring values and interests. This "self-knowledge" can also be regarded as a virtue and it too, must be developed. Not only does one need to ask of oneself the right questions, but one also needs to be exposed to a robust domain of activities in order to discover those that best align with one's enduring values and interests. Self-concordance does not just happen, but must be developed by cultivating the various competencies on which it depends.

Aristotle's Approach Modernized

The modern virtues

Having reviewed some of the preconditions of self-concordance, we are now in a position to reflect on how identifying flourishing as self-concordance captures some of the fundamental tenets of Aristotle's ethical theory. As should be clear, the notion of self-concordance provides the *telos* necessary to ground a modern theory of the virtues. The virtues are simply the conditions necessary for self-concordance. Again, self-concordance is a specific type of teleological activity that expresses the core values and interests of the individual. Engaging in this activity requires both knowledge of one's enduring values and interests, *and* the aforementioned capacities underlying the executive function. Knowledge of one's fundamental values and interests provides one with the *specific* ends at which one should aim, the specific content of one's "calling." But again, knowledge of this content is not sufficient for self-concordance, for one also needs the various analytic, navigational, and motivational competencies that enable one to realize one's calling through action. And these various competencies, as well as the knowledge about one's fundamental values and interests, are *virtues*. Aristotle claims that a virtue is what makes the work or *ergon* of a human being be done well (*NE* 1106a). If we understand the *ergon* of a human being as self-concordance, then self-knowledge and the perfection of the analytic, navigational, and motivational competencies are virtues.

Importantly, the *kind* of virtue theory that develops pursuant to a view of flourishing *qua* self-concordance aligns closely with the kind of virtue theory developed by Aristotle. As I argued earlier, Aristotelian virtue is very different from the Judaeo-Christian conception of virtue in which "other regard" is primary. On the Judaeo-Christian conception, the term "virtue" has connotations of altruism, charity, and benevolence. For Aristotle, however, the term "virtue" is much more narrow. In Aristotle's view, the virtues are attributes of the motivational system that enable its possessor to engage in the actions that lead to contemplative excellence. Aristotle's virtue theory, therefore, focuses more on individual development and flourishing than it does on "communal harmony" or social benevolence. In like manner, identifying flourishing as self-concordance allows us to construct a modern theory of virtue primarily focused on the attributes of the individual's motivational system that enable him to engage in self-concordant activity. This, of course, does not mean that communal harmony is unimportant. Self-determination theory, recall, identifies *relatedness* as a fundamental human need.[55] A modern virtue theory which identifies optimal fulfillment of human needs as virtues, can easily include the other-regarding competencies required to optimally satisfy the need for relatedness.[56]

The awakening of proper pleasure

Yet another similarity between SDT and Aristotle's view of flourishing concerns the role of proper pleasure. We saw in Chapter 2 that Aristotle identifies flourishing with the activity of a motivational system that is commanded by the proper pleasures. Thus, in Aristotle's view, the correct affective content of the ultimate desire—the calling—that drives a human being's motivational system concerns rational activities since those produce the proper pleasures.

As we have learned, self-concordant activity is a special type of intrinsically motivated pleasurable activity insofar as it aligns with the individual's enduring values and interests *and* provides him with the pleasurable perception that his actions flow from himself and not from another. If the pleasure of self-concordant activity is regarded as a proper pleasure, then flourishing can be understood as the activity of a motivational system that is commanded by proper pleasure. In Chapter 2, we noted that Aristotle identifies flourishing with a specific motivational system that is driven by the proper pleasurable activities, and hence by the *correct* calling. The same holds true for a modern view that identifies flourishing with self-concordant activity. The correct calling on the modern view consists in the self-concordant activities of the individual. Again it is the specific pleasure—the specific "hedonic cocktail" of self-concordant activities that should drive and command the individual's actions.

Understanding that the specific "hedonic cocktail" characteristic of self-concordant activity is a proper pleasure, helps us to see yet another similarity between Aristotle's theory and a view in which flourishing is identified as self-concordance. For Aristotle, proper pleasure must be *awakened, stimulated and nurtured*, so that it, rather than some other pleasure, commands the motivational system. And it is through habituation by which one awakens this proper pleasure. Remember, habituation is a correction of the motivational system. Human beings are not born with the virtues and therefore the default motivational system is commanded by the *wrong* hedonic cocktail.

Again, a completely "naturalistic" explanation for this is that given that one's repeated actions leave a mark on one's character (*NE* 1114a), and given that all of one's actions for the first few years of life are impulse driven, then the resulting motivational system is one that is driven by the bodily pleasures. By repeatedly engaging in actions directed by practical thought, *despite* the fact that desire pulls in the other direction, one gradually corrects the motivational system. Why? I have argued that it is because repeated compliance with practical reason introduces a proper pleasure into the motivational system, that, given human nature, the trainee must prefer. As we shall see, the fundamentals of this story remain the same on a modern view which identifies self-concordance as a crucial component of flourishing.

Self-concordance is *intrinsically motivated* activity that aligns with one's enduring values and interests. But for much of one's life, one's actions are *extrinsically* motivated. School, which occupies a significant portion of the lives of most people's early years, is replete with activities that issue from another. Given that extrinsically motivated actions are ubiquitous for much of one's life, the laws of habit dictate that our character is affected accordingly. And the result is a motivational system that has an habitual "extrinsic motivation" orientation. We become accustomed to acting in accordance with goals that are not our own, and to seeing ourselves act in this way. Indeed, this is our default setting and it is responsible for a kind of passivity that is incompatible with self-concordant activity.

Complicating the matter further is the fact the default "interpretive system"—which serves the motivational system—is also one of passivity. The motivational system is cued by our perceptions, judgments and interpretations, the "interpretive system." But the default setting of our interpretative system is one of naiveté. We must *learn* not only that we are responsible for our interpretations of things, but also that we can re-engineer our judgments and causal attributions so that they contribute to self-concordance. Just as our default motivational system contributes to a kind of passivity that is incompatible with self-concordant activity, so too does the default interpretive system. Both default systems must be overcome in order to realize self-concordance.

The fact is, one who has not overcome these default settings has not tasted the proper pleasure—the specific hedonic signature associated with self-concordant activity. Consequently, our default motivational system is one that is commanded by the *wrong affect.* Instead of being commanded by the proper pleasure associated with a very special type of intrinsically motivated activity, a human being's default motivational system tends to be commanded by other pleasures, such as the pleasures associated with immediate impulse. However, these pleasures are not the ones that *should* be commanding the motivational system since they do not and cannot satisfy one's fundamental need for self-concordant activity.

If the default motivational system is commanded by the wrong affect, a correction is needed. And the correction that is needed proceeds in virtually the same way as Aristotle describes the development of affective virtue. That is, to correct the default motivational system, one repeatedly engages in the steering function of private speech on interfering habit triggers both external and internal. For it is by doing this, as we know, whereby one develops habitual self-regulation, the *sine qua non* of self-concordance.

When one first uses private speech in non-play contexts, one must consciously and deliberately engage in an act of will power. There must be moments, that, when faced with a temptation that thwarts one from the goal, and/or when faced with an unhealthy causal attribution, one consciously and deliberately steers her attention accordingly. No one can do this but the individual herself. I submit that this repeated act of deliberate mindfulness awakens the proper pleasure associated with self-concordant activity. Why? Because by engaging in an act of deliberate mindfulness, *one begins to perceive one's own voluntariness and one's own agency,* and *that,* given a human being's fundamental need for autonomy, is pleasurable. I submit that to voluntarily engage in the steering function of private speech, *whether it is used to overcome an external or an internal habit trigger, and/or to induce motivation,* to succeed in this act, and to see oneself doing so, has a specific "hedonic signature" that, if repeated enough times, awakens the individual to more pleasurable forms of action.

I am by no means claiming that the deliberate use of the steering function of private speech is sufficient for self-concordance.

As mentioned above, a *number* of virtues are needed for self-concordance, including the self-knowledge of one's enduring interests. But the deliberate use of the steering function of private speech is necessary not only because it introduces the individual to more pleasurable forms of action, but also because self-regulation and other important competencies are developed by the repetition of such actions.

In the following chapter, I briefly discuss the role that the legal/political ecology should play in the cultivation of the modern virtues. To what extent should the legal/political ecology nurture the human capacities underlying a human being's ability to set and realize ends of one's choosing? Aristotle believed that the polity should play a role in the cultivation of flourishing, and so a modern Aristotelian approach is one that acknowledges such a role played by the legal/political ecology. Although defenders of liberalism cringe at the prospect of state inculcated virtue, the modern virtues described herein are not incompatible with autonomy. In fact, they are essential for it. A regime that avoids providing ecologies that cultivate the inner resources involved in the setting and realization of one's own ends is one that makes it more difficult for its citizens to be autonomous and hence to flourish. In the concluding chapter, we will see that there are a number of institutions that are creating ecologies aimed at cultivating the inner resources that enable individuals to set and realize self-concordant ends.

Chapter 6

The Ecological Preconditions of Self-Concordance

Introduction

In this chapter, I briefly discuss how understanding flourishing as self-concordance harmonizes in with Aristotle's view about the function and purpose of law. In chapter 4, I used the term "normative ecology" to refer to the normative system (the collection of political and legal norms) which helped to cultivate the virtues of Aristotle's students. And while it is true that, of Aristotle's ideal state, the normative ecology included both moral norms (moral ecology) and legal norms (legal ecology)—to the point that the "moral ecology" was virtually co-extensive with the "legal ecology," this is untrue in the state that embraces contemporary liberalism. In the liberal state, the distinction between moral and legal norms is greater: there is less overlap between moral norms and legal norms. True, there may be overlaps between the moral and legal ecologies within the liberal state, but these overlaps are much more narrow than they were in Aristotle's day. The diagram on the following page depicts the differences between the normative ecology of Aristotle's ideal state, and the normative ecology of contemporary liberalism.

Aristotle believed that the primary aim of law was to promote human flourishing by commanding the habits that produce the necessary "inner resources" and hence virtues (*NE* 1102a, 1180a). Although this view of the political community is regarded as incompatible with liberalism,[1] I will show that it is not. Essentially, I argue for a view that both embraces Aristotle's conception of normative ecology and also recognizes the centrality of individual autonomy—the chief value of the liberal state. To supplement this discussion, I will discuss a number of existing programs that cultivate some of the

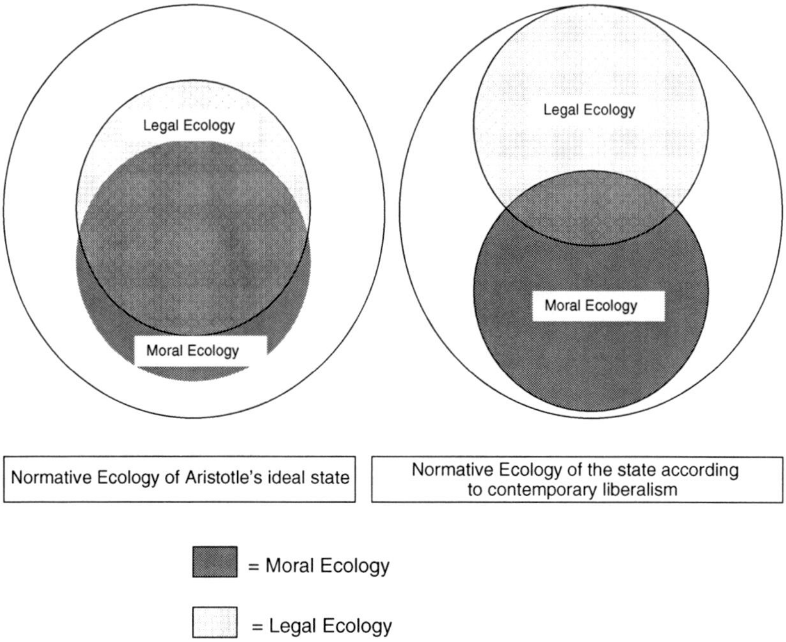

preconditions for self-concordance. Because they are "external" to the individual, these programs can be thought of as components of the individual's "ecology" or surrounding environment.[2] As we shall see, the "political/legal ecology" discussed herein is not confined to notions of federal or constitutional law, but is much more broad. References to the "state" refer to the notion of *political community or polity* and therefore include local governmental units such as municipalities that shape shape and control policy, especially at the educational level. I use the term "legal ecology" to refer to this broad notion of the state.

The role of legal ecology

The analytic, navigational and motivational competencies that are necessary for self-concordance are, of course, *internal* to the self insofar as they pertain to the inner resources of the individual. But Aristotle believed that the legal ecology—which is external to the individual—is essential in promoting flourishing (*NE* 1179b). For the legal ecology inculcates the habits necessary for the cultivation of the inner resources necessary for flourishing. And although he recognizes that the individual is largely responsible in acquiring these inner resources, Aristotle denies that the individual should be alone in her effort. In his view, one should have help from the law and from education. In Aristotle's view, the political community exists in order to assist the individual in developing the inner resources that are necessary for flourishing. Law and education provide the *external or ecological preconditions* that cultivate and preserve the requisite inner resources (*NE* 1180a). Aristotle, therefore, implies that the state has a substantive positive duty, namely, to provide a legal ecology that cultivates the inner resources which enable the individual to flourish. Through the power of law and education, individuals are prodded to repeatedly perform the actions that, over time, give rise to the motivational system needed for flourishing.

Although many regard this view of political community and of law as incompatible with liberal conceptions of "the good," I hope to show that it is not. If we discard the assumption that autonomy is simply the "absence of constraints," and replace it with the notion that autonomy is the *presence* of a specific kind of teleological activity (self-concordant activity), then we can understand the "virtues" as the preconditions necessary for this activity. A polity that provides the ecology that inculcates these virtues is not one that is antithetical to autonomy, but one that nurtures, cultivates, and grows it, providing the "earth which nourishes the seed" (*NE* 1179b).

The fact is, self-concordance requires a specific legal ecology. The actualization of one's potential for self-concordance does not happen willy-nilly, but only if certain skills are developed and proper pleasures are awakened. Although the individual is ultimately responsible for the development and awakening of these inner resources, the actualization of these inner resources largely depends upon a specific ecology. We have said that self-concordance requires that the individual is exposed to a healthy variety of possible ends and activities so that she can find those that best align with her enduring values and interests. Thus, self-concordance requires a robust "marketplace of ends" from which the self chooses (or perhaps is chosen by) the appropriate activities. *Self-concordance requires that the individual is exposed to a thriving ecology of possible selves.* Thus, self-concordance requires a certain legal ecology which respects the boundaries of the self and recognizes the rights of individuals to shape their destiny. One way that this can be achieved is by granting privacy rights that create a "zone of privacy" which provides the existential space that the self needs in order to shape its destiny and actualize its potential for self-concordance. As Justice Kennedy writes in the landmark decision *Lawrence vs. Texas*, "Liberty presumes an autonomy of self that includes freedom of thought, belief, expression, and certain intimate conduct."[3] Respecting liberty implies that the state recognizes a "zone of privacy"—an ecological space with which the state cannot unreasonably interfere by, for example, circumscribing the marketplace of ends and limiting the possibilities for the self. But Aristotle rejected the idea that all a polity needs to do is "stand back" and refrain from interfering. The view that the political community only has negative duties towards its citizens is a radical departure from Aristotle's view. Aristotle believed that the primary duty of the political community was not to refrain from interfering but to cultivate and nurture the habits and virtues necessary for flourishing. An ecology of non-interference is necessary but not sufficient for flourishing. Flourishing also requires an *ecology of nurture*, and hence a legal ecology that assists the individual in developing the virtues associated with teleological excellence.

The rise of secularism in conjunction with liberal theories of the good has led to a rejection of views such as Aristotle's which impose upon the state a positive duty to cultivate specific virtues. Proponents of liberalism believe that the citizens should be free to pursue their *own* conceptions of the good, and hence that the political community (via public education and law) should refrain from imposing any substantive conception of the good onto its citizens. Since liberalism rejects the idea that the state should impose a substantive conception of the good onto its citizens, it seems to depart from Aristotle's view that the political community has a duty to promote the inner resources and virtues necessary for flourishing.

However, if flourishing is identified with self-concordance, then it is possible to combine Aristotle's view with liberalism. Self-concordance is a conception of the good, but is a "thin," open-textured good that can be instantiated in an infinite number of ways. But again, self-concordance does not arise naturally, but must be cultivated and developed. As we have seen, certain competencies (self-regulation, the steering function of private speech, resilience, hope, and self-efficacy, etc.) must be present in order for self-concordance to be actualized and fully manifest. Thus, although a system aimed at cultivating, promoting, and protecting self-concordance implies an ecology—policies and programs—aimed at cultivating a number of specific "virtues," these virtues are preconditions of an open textured notion of the good that is completely consistent with liberalism. Imposing a conception of good *qua* self-concordance onto citizens through law and policy is not inconsistent with the sentiment that citizens should be free to pursue their own conceptions of the good. In fact, the view of flourishing developed here actually supplements liberalism insofar as it recognizes that *a certain motivational system must be present if an individual is to realize her own conception of the good,* and that *such a motivational system is developed by cultivating the inner resources on which it depends.* However, following Aristotle, the state has a duty to assist the individual in cultivating these inner resources. Again, by providing an ecology that supports self-concordance, the polity functions as the "earth which nourishes the seed" (*NE* 1179b).

Programs that cultivate

Today, a handful of educational institutions have implemented programs aimed at cultivating some of the capacities underlying effective goal attainment. Additionally, there are programs that help nurture an individual's ability to discover the ends and activities that can serve as her self-concordant goals. The ability to discover these activities, of course, requires that the individual is exposed to a robust range of possible ends and activities. The innovative after-school program, *Strong Women, Strong Girls,* founded by Lindsay Hyde,[4] is designed to develop the self-efficacy and self-esteem of girls by exposing them to possible future selves by combining stories about inspiring, professional women with concrete skill building activities. Broadening one's conception of what is possible for oneself is a critical first step in the recognition and adoption of self-concordant goals.

As we will soon discuss, a novel substance abuse program known as the Brooklyn Program, is premised on the belief that developing a healthy conception of one's "possible self" is essential to rehabilitation and recovery.[5]

Other programs cultivate some of the different competencies underlying goal attainment. For instance, the Knowledge is Power Program (KIPP) network of schools that was created by David Levin and Michael Feinberg focuses on cultivating the traits of self-control, hope, and self-efficacy.[6] The success of this program has won its creators 23 million dollars in grants, 8 million of which is coming from the Bill and Melinda Gates Foundation.[7] The "Self-Science" curriculum used at the private Nueva school in California, moreover, includes a unit wherein the student is asked to reflect on her learning habits.[8] For example, the student is asked to reflect on her learning style and to make explicit the times of day as well as the methods which both facilitate and hinder her learning. Such an activity cultivates the mindfulness necessary to engage in the various forms of masterful self-regulation discussed in Chapter 5.

A preschool program focused on cultivating both the analytic and navigational components underlying goal attainment is the "Tools of the Mind" program, developed by Elena Bodrova and Deborah Leong.[9] The curriculum of the *Tools* program includes "play plans" which require the student to strategically think about her play time. Immediately before play time begins, children are asked to write down and draw what they plan to do during play, and to determine the props and roles needed for their plans.[10] At the end of the play period, children are asked if they want to continue the play in the next play period, and if so, they determine the props needed to continue the play.[11] Such an activity is aimed at the developing the analytic competencies and planning abilities underlying goal attainment. But the *Tools* program also focuses on the development of the navigational competencies. The program does this by focusing on the cultivation of self-regulation through a variety of games and opportunities for "free play." As we discussed earlier, free play activities are considered to be the first opportunities in which the child self-imposes constraints on her behavior, and hence begins the development of self-regulation.[12] The *Tools* program acknowledges the developmental link between free play and self-regulation and hence incorporates free play into its curriculum.

Some legal interventions are beginning to focus on cultivating the preconditions of self-concordance. Consider a novel substance abuse program, the "Brooklyn Program," which oversees the court mandated substance abuse treatment of federal offenders. This program is premised on the view that, to be adequately treated, the authenticity and self-actualization of the offender must be cultivated:

> By directing his efforts towards future behavioral change in areas implied by . . . self-generated directions, the substance abuser or addict can come to a fuller, more positive and rewarding answer to the questions of life and so . . . begin to choose to leave the problem behavior behind . . . the program consists of a series of exercises designed to create a deepened sense of Self and personal direction by assembling successive layers of positive experience into deeper, more global and more accessible approximations of a core identity with the direction implied thereby.[13]

With the cultivation of self-concordance as its ultimate goal, the Brooklyn Program aims to cultivate the inner resources required for self-concordance, including hope and self-efficacy:

> [There are] . . . three necessary elements in substance abuse treatment: Self-efficacy, Futurity, and Self-esteem . . . self-efficacy . . . holds that people need to have experiences of success in order to attempt a task, find the motivation to continue in a task, and feel good about themselves in the context of that task . . . self-esteem refers to feelings of positive self-regard that result from experiences of efficacy in multiple activities . . . Futurity, as applied here, entails the discovery of goals and activities that are inherently meaningful to the offender. It is, in many cases, the discovery of a life goal or spiritual mission that provides the appropriate impetus to change.[14]

To cultivate self-efficacy, futurity, and self-esteem, participants in the Brooklyn program are asked to reflect upon positive memories with respect to five different categories: focused attention, good decision-making, a moment of discovery, fun, and confidence in a practiced skill.[15] These positive memories form the fundamental building blocks of a concrete representation of a future self to which the participant can then aspire:

> the [memories] are assembled into a single complex state that we understand to be a constellation of a deeper sense of Self . . . the unrealized whole towards which healthy personal development strives. While the individual [memories] are useful as building blocks, their capacity to assemble a much deeper and continuing sense of this Self provides more permanent and enduring changes. It is in itself a resource state, but it also begins to awaken the individual to his or her identity with a continuing Self who can transcend the momentary vagaries of existence.[16]

The Brooklyn program is novel and relatively new, having begun in the late 1990s.[17] In terms of efficacy, 55% of the participants remain abstinent after leaving the program,[18] and the program is known as one which "is characterized by high rates of retention and low relapse rates."[19]

The Brooklyn program is an illustration of a legal intervention aimed at self-concordance. But such interventions need not be restricted to substance abuse treatment. One can also imagine interventions within a wide variety of contexts including prison populations. However, if the incarceration period of the prisoner is sufficiently long, this could indeed frustrate the development of the hope and self-efficacy needed for self-concordance. At any rate, legal interventions are just beginning to aim at cultivating self-concordance, and the Brooklyn program can certainly be used as a model for other reforms which adopt flourishing *qua* self-concordance as their ultimate end. *Qua telos*, self-concordance can animate and direct meaningful reform.

But programs aimed at cultivating the competencies necessary for self-concordance need not be limited to children, addicts, or inmates. The success of the executive or life coaching movement in which "life mapping" figures prominently testifies to the fact that many individuals seek help in achieving their goals.[20] A life map is simply a personal strategic plan which typically includes a vision and an action plan for realizing the vision. There are a variety of approaches to life mapping, some more rigorous than others.[21] These approaches, however, tend to stress the analytic component of the executive function as they focus on how to analyze goals into more simple components. But as we have seen, navigational and motivational competencies are just as important.

A variety of programs exist to cultivate the various competencies underlying self-concordance. The fact is, however, few (if any) policies have consciously and explicitly embraced the concept of self-concordance as a regulative ideal. *Qua telos* and ultimate end of action and of policy, the notion of self-concordance can provide structure and direction for innovative programs that are focused on developing its underlying competencies.

Conclusion

As I have shown, by harnessing the view of human flourishing that is found within self-determination theory, a modern virtue theory that embraces the fundamental tenets of Aristotle's ethics can be constructed. On this view, the proper role of the polity is not to abstain from the lives of the citizens. On the contrary, the state, through education, policy and legal interventions, has a duty to provide both a sphere of liberty that enables the individual to choose the specific constellation of self-concordant goals in which her conception of the good consists, *and* to cultivate the virtues—the preconditions—necessary for self-concordance. Indeed, it is *because* flourishing consists in autonomy *qua* self-concordance, that the political community thereby has a duty to secure the conditions, both external and internal, necessary for its realization.

As we have seen, Aristotle recognized the polity's duty to cultivate the inner resources necessary for flourishing. And as we have also seen, these inner resources can be thought of as "moral" virtues.

I have argued for a conception of virtue in which the virtues are identified with the various competencies (analytic, navigational, and motivational), that are necessary for self-concordant activity. Legal scholarship has been slow to realize the role of law and policy in cultivating these inner resources. In one recent article, two scholars note that the cultivation of affect is "a dimension of the law yet to be explored."[22] But one reason *why* this dimension of law should be explored, is that it enables us to accommodate the modern notion that autonomy is sacrosanct, with the ancient idea that the ultimate end of law is to promote the virtues essential to human flourishing. Synthesizing a developmentalist interpretation of the *NE* with a modern view of flourishing provides a new paradigm for education and law that captures Aristotle's deepest convictions about both the nature of flourishing and the state's role in promoting it. On the modern view, the *telos* is self-concordance both for the individual *and* for the polity, and it is incumbent upon both to engage in actions that cultivate the inner resources by which self-concordance is actualized. In the liberal state, the gulf between the moral and the legal ecology need not be so wide. Indeed in the liberal state, there should be a broad overlap between the moral and the legal ecology. This echoes Aristotle's statement that the end of the best man and the best constitution is one and the same (*Pol.* 1324a). But here, the best man is the self-concordant man and the best constitution—the best legal ecology—is one that nurtures the inner resources that lead to self-concordant activity.

Aristotle, we should note, recognized the importance of role models—of specific individuals that one could point to within one's community that demonstrated the virtues. He writes, "regarding practical wisdom we shall get at the truth by considering who are the persons we credit with it" (*NE* 1140a). And as an example, he mentions Pericles (*NE* 1140b). Notably he also states that philosophers such as Anaxagoras and Thales did *not* have it (*NE* 1141b).

I believe that embracing self-concordance as a regulative ideal requires that we, too, can point to specific individuals who exhibit this masterful teleological ability. And so to that end, I conclude this book by pointing to one specific model who exemplifies self-concordance, Wendy Kopp.[23]

Wendy Kopp: An Illustration of Self-Concordance

Wendy Kopp is the founder of the organization *Teach for America*,[24] and she exemplifies self-concordance. Kopp not only understands what her own enduring values and interests are, but she also possesses the competencies required to engage these interests in specific activities, masterfully and beautifully.

The mission of *Teach for America* is the eradication of educational inequality. As Kopp explains in her book *One Day All Children,* she first learned about the issue first hand when she witnessed the difficulties encountered by her college roommate at Princeton University. Unlike some of the other Princeton students who attended elite east coast prep schools, Kopp's roommate attended the public schools of the South Bronx.[25] Kopp knew some of these prep-schoolers and they described Princeton as a "cake walk." Thus, Kopp bore witness to the fact that success at Princeton was a function of one's prior educational experiences: for some it was grueling, for others it was a "cake walk." This fact could have easily evaporated like so many other experiences, but for Kopp, it did not. It stuck. It struck a chord and deeply resonated with one of Kopp's enduring values regarding the importance of education.

Kopp was so bothered by the divergent experiences of her peers, that she organized a conference on the problem with like minded students and colleagues. Here we see the first display of an intrinsically motivated action sparked by Kopp's bothersome experience. She did not organize this conference for a class, but because educational opportunity was something that she deeply valued. It was while at the conference that she had the idea for a National Teacher Corps. Naturally, she devoted her senior thesis to the topic. Note how her activities flow from and capture her enduring values and interests. The witnessing of her struggling roommate juxtaposed with the breezy attitude of the prep-schoolers compelled Kopp to engage with this issue through action and personal reflection. Quite literally, it was her value in education driving her, animating her actions, and directing her decisions.

After deepening her learning about educational inequality, a self-concordant goal was born. The goal: *the elimination of educational inequality*. The means? *Teach for America*. As Kopp explains in her book, the deep learning and reflection that was the result of her senior project compelled her to create not just an organization *but a movement* devoted to improving the educational opportunities for all children. The organization *Teach for America* was envisaged by Kopp to be the mechanism that creates this movement. She thought, *if my goal is the elimination of inequality, then I need a movement to deal with this huge problem, and in order to create this movement, I need an organization.* Thus, Kopp articulated to herself the broad contours of her strategic plan for her ultimate goal. This is the analytic competency at work. Note the sequence of hypothetical necessities in Kopp's plan: Teach for America is both a goal to be obtained *and* a means to some further goal, that is, a national movement. But neither a national movement nor *Teach for America* is the most final goal of Kopp's plan. Both are *goals* and *means* that exist for the sake Kopp's ultimate goal of eradicating educational equality. It is the securing of *this* value that provides the content of *Kopp's* calling.

Kopp did not merely chart out the path for the attainment of her goal. She acted. She engaged her body in a sequence of actions that were performed for the sake of realizing her long-term vision: letter writing, fundraising, team-building, and so on. This manifests the navigational and motivational competencies at work. Kopp was also able to overcome numerous potential derailments along the way: lack of funding, an unfair press, etc. She was never derailed. Indeed, she is still going. At the end of her book she writes, "[k]nowing that our vision is within reach, my colleagues and I are more determined than ever to do whatever it takes to get there."[26] Aristotle's claim in the *NE* that the pleasures of thinking and learning "will make us think and learn all the more" is apt. What Kopp's story bears witness to is the fact that the pleasures of self-concordant activity intensify and strengthen this activity.

It is important to point out that Kopp's genius is not merely in her goal, but also in the masterful navigational and motivational competencies. Discovering self-concordant goals is the (relatively) easy part, *doing them, achieving them* is the hard part. Because for many, the necessary correction of the default motivational system has *not* been made, not only do most people find it difficult to get started (as this inevitably involves discomfort), but most people also find it difficult to stay on course when confronted with failure, fatigue, frustration, temptation, discouragement, and so on. Kopp exhibits the necessary competencies, beautifully and masterfully. Individuals like Wendy Kopp are the *phronimoi* of our time. And although contemplating and reflecting on Kopp and other modern *phronimoi* may provide a source of pleasure, much work needs to be done in creating ecologies—both private and public—that nurture self-concordance and help to grow such individuals. Therefore, let us not forget Aristotle's emphasis on action when he says:

> *Where there are things to be done the end is not to survey and recognize the various things, but rather to do them.* (*NE* 1179a)

Notes

Chapter 1

[1] In the film, the character Peter Gibbons considers what he would do if he had a million dollars and therefore did not have to work. His response is "I would relax, I would sit on my ass all day, I would do nothing." The implication here is that Gibbons' desire for flourishing would be satisfied by "doing nothing."

[2] See Cooper (1975), Devereux (1981), Hardie (1965), Kenny (1978), and Kraut (1989).

[3] See Ackrill (1980), Irwin (1991), Keyt (1983), and Roche (1988).

[4] Cooper (1975), Whiting (1986), Irwin ([1988], 363–72).

[5] Hardie ([1965], 279).

[6] Translations from the *Nicomachean Ethics* are from David Ross' Oxford Translation, reissued as an Oxford World Classics paperback, revised by J. L. Ackrill and J. O. Urmson (NY: Oxford University Press, 1998). An older version of this translation is available on the free internet at http://classics.mit.edu/Aristotle/nicomachaen.html [sic]

[7] Sartre ([1957], 15).

[8] Irwin ([1980], 35–6).

[9] An exception is Whiting (1986) who argues that the *idion* of man is identical to man's essence. However, Ackrill, for instance, writes, "Aristotle has clearly stated that the principle of the *ergon* argument is that one must ask what powers and activities are peculiar to and distinctive of man." Ackrill ([1980], 27). And similarly, Richard Kraut, "[o]ur function cannot simply be to live, since that is something we have in common with plants, and we are looking for what is peculiar—to human beings . . . [Aristotle] . . . is looking for what is peculiar (*idion*) to human beings." See Kraut ([1989], 313).

[10] See references in Roche ([1988], 192, n.31). Textual evidence seems to rule out such an explanation. For in Book X of the *NE*, Aristotle seems to refer back to his discussion of flourishing in Book I, wherein he tells us that flourishing is a *teleiotatos*, self-sufficient, and intrinsically desirable end (*NE* 1097a–1097b); and that flourishing is an activity in accordance with the best and *teleiotatos* virtue (*NE* 1098b). In chapter X.7, these remarks are echoed. There, Aristotle states that *sophia* is the highest (*kratistos*) virtue, that

---

Visit http://www.aristotlesethics.com for easy access to the web links mentioned in the notes.

it is the best (*aristos*) thing in us, and that activity in accordance with *sophia* is *teleia* happiness (*NE* 1177a). Although Aristotle never explicitly claims that *sophia* is identical to the *teleiotatos* end, he seems to imply this much. For Aristotle's reference to *sophia* as *aristos* seems to refer back to the conclusion of the function argument, which states that *eudaimonia* is an activity of the soul in accordance with the best (*aristos*) and *teleiotatos* virtue (*NE* 1098a). Moreover, Aristotle states that contemplative activity is self-sufficient (*autarkeia*), and desired for its own sake (*NE* 1177a–b); and in Book I, he tells us that these two qualities are characteristic of *eudaimonia* (*NE* 1097a–b). Thus, Aristotle seems well aware of his preceding discussion about *eudaimonia*, even in the final book of the *NE*.

11 Hardie ([1965], 279) and Cooper ([1975], 165–80). Cooper argues that Aristotle's inconsistency in the *NE* is caused by the fact that, in the earlier books of the *NE*, Aristotle accepts the psychological theory of the *Eudemian Ethics*, whereas at other points, Aristotle accepts the psychological theory of the *De Anima*. Since these two theories are inconsistent with one another—the psychological theory of the *Eudemian Ethics* identifies a human being as a combination of mind and body, but the psychological theory of the *De Anima* identifies a human being with *nous*, which is non-bodily—Aristotle's discussion of the human good is inconsistent. Cooper (1987) has rejected this rather uncharitable reading and now embraces inclusivism.

12 See Cooper (1987), Keyt (1983), Kraut (1989), Purinton (1998), Roche (1988), and Whiting (1986).

13 See Cooper (1975), Devereux (1981), Hardie (1965), Kenny (1978), and Kraut (1989). Kraut claims that ethical virtue is an intrinsically desirable good, but that it is not a component of *eudaimonia*. See Kraut ([1989], 215). Irwin (1991) has persuasively argued that Kraut's interpretation is inconsistent with Aristotle's *eudaimonist* project. For Kraut claims that ethical virtue is intrinsically *and* instrumentally desirable. Irwin points out that this is problematic. For why should one choose ethical virtue for its own sake, if it is not a component of *eudaimonia*?

14 See Cooper ([1975], 149–68), Kraut ([1989], 181).

15 This is consequence of Kraut's claim that the more theoretical activity that one engages in, the more *eudaimon* one is. See Kraut ([1989], 9). See also Cooper ([1975], 149–50).

16 Kraut ([1989], 181).

17 Inclusivists also reject intellectualism because proponents of this view typically claim that ethically virtuous activity is merely instrumentally valuable. See Cooper ([1975], 163–5), and Kraut ([1989], 5–6). Kraut argues that even though ethical virtue is not a component of the most happy life, it is nevertheless both instrumentally and intrinsically desirable. However, Irwin ([1991], 383–86) points out that the claim that the *eudaimon* individual values certain intrinsic goods that are *not* components of *eudaimonia* is incompatible with Aristotle's eudaimonistic project. But there is a further problem for the view that ethical virtue is merely instrumentally valuable. For if it is true that the virtues of character are valuable only when being used to help promote theoretical activity, then it follows that since courage is one of the virtues of

character, then courageous activity is only valuable if it promotes theoretical activity. But this is absurd. For Aristotle defines courage as "being fearless in the face of death in battle" (*NE* 1115a), and it is rather difficult to see how such a virtue could help promote theoretical activity.

18 Roche ([1988], 191). Although all inclusivists agree that *eudaimonia* is an end that contains a plurality of virtues, they disagree about which goods comprise *eudaimonia*. Ackrill (1980) claims that all intrinsically desirable goods comprise *eudaimonia*. Purinton (1998) disagrees, claiming instead that all and only virtuous activities, activities of the rational soul, comprise *eudaimonia*. Keyt (1983) holds a similar view to that of Purinton's, claiming that all of the rational virtues comprise *eudaimonia*.

19 Ackrill ([1980], 19–20).

20 Ackrill (1980) suggests that theoretical and ethically virtuous activity are equally desirable.

21 See Irwin (1991), Keyt (1983), and Roche (1988) for the view that theoretical activity is more desirable than ethically virtuous activity. Importantly, although moderate inclusivists claim that theoretical activity is more valuable than activity that is ethically virtuous, they do not also believe that one should eschew the ethical virtues in order to maximize theoretical activity, as do the intellectualists. For moderate inclusivists claim that ethical virtue is a necessary condition of theoretical activity. As David Keyt puts it, ". . . theoretical activity is to be maximized but only within the constraints of practical wisdom and moral virtue. Moral activity is the foundation, and theoretical activity the superstructure, in the best life for a man." Keyt ([1983], 370). On such a view, it is impossible for ethical virtue to be compromised for the sake of theoretical activity, since ethical virtue is necessary for its operation.

Chapter 2

1 Hume ([1748], 41–52).

2 All translations of *De Anima* are by Hugh Lawson-Tancred (1986), unless otherwise noted.

3 Sartre ([1957], 15).

Chapter 3

1 David Keyt ([1983], 370) claims that ethical virtue provides a necessary foundation for contemplative activity and therefore the latter cannot be severed from the former. But Keyt does not explain how ethical virtue is a foundation of theoretical activity, nor does he explain why Aristotle holds such a view. I like to think of my view as a supplement to Keyt's. I offer an explanation grounded in Aristotle's biology, as well as his psychology and epistemology, that explains the precise way in which ethical virtue provides a foundation for theoretical activity.

[2] See Davidson (1990) and (1995) for a discussion about the different neural substrates underlying "approach" and "avoidance" behavior. Davidson's research demonstrates that activity in the left anterior region of the brain is connected with the positive emotions associated with approach oriented behavior, whereas activity in the right anterior region of the brain is connected with painful emotions associated with withdrawal/avoidance oriented behavior.

[3] James ([1899], 42). The full text of *Talks to Teachers* is available on the free internet at: http://www.des.emory.edu/mfp/james.html#talks. The passage quoted can be found in Chapter 8, entitled "The Laws of Habit" *available at*: http://www.des.emory.edu/mfp/tt8.html

[4] The imagery of the frozen sea and the axe is from Kolak and Martin ([1999], 1). There, they claim that "There is a frozen sea within us. Philosophy is an axe." As an undergraduate at William Paterson College, I had the great fortune and privilege to learn from this book and from Professor Kolak. Of course, if philosophy is the axe to use upon the sea within, one must lift the axe and strike. Not just once, not just twice, but repeatedly.

[5] *See* Kandel ([2006], 261–276).

[6] *Id.*

[7] See Reeve ([1995], 34–45).

[8] See Cooper ([1975], 149–68), Kraut ([1989], 181).

[9] *Id.*

[10] *Id.*

[11] *Id.*

Chapter 4

[1] *Butchers' Union Co. v. Crescent City Co.*, 111 U.S. 746 (1884). *Available at,* http://caselaw.lp.findlaw.com/scripts/getcase.pl?court=US&vol=111&invol=746

[2] "[C]ertain inherent rights lie at the foundation of all action, and upon a recognition of them alone can free institutions be maintained . . . Among these inalienable rights . . . is the right of men to pursue their happiness, by which is meant the right to pursue any lawful business or vocation, in any manner not inconsistent with the equal rights of others, which may increase their prosperity or develop their faculties, so as to give to them their highest enjoyment." *Id.* at 757.

[3] Foot (2002), Anscombe (1958), MacIntyre (1984).

[4] Farrelly & Solum (2008).

[5] Annas ([2005], 526–7).

[6] See Little, et al. (2007) for an overview of an interesting and important branch of psychology known as "personal project analysis".

[7] Gable & Haidt ([2005], 104).

[8] Sheldon & King ([2001], 216)

[9] *Id.*

[10] Gable & Haidt, *supra* n. 7, at 103.

[11] *Id.* at 104.

[12] *See* Csikszentmihalyi (1990) and (2000). For more information on Csikszentmihalyi's concept of flow, *see* http://en.wikipedia.org/wiki/Flow_(psychology)

[13] Csikszentmihalyi ([1990], 4).

[14] *See,* e.g., Diener, et al. (1998).

[15] Ryan & Deci (2001).

[16] *Id.* at 145.

[17] *Id.*

[18] For an overview of Maslow's hierarchy of needs see, http://en.wikipedia.org/wiki/Maslow's_hierarchy_of_needs

[19] For an excellent survey of this research, see Deci & Ryan (2002).

[20] Ryan & Deci (2000).

[21] Sheldon & Elliot (1999).

[22] Deci & Ryan ([2002a], 5).

[23] White (1959).

[24] Thomas Hobbes champions this conception of autonomy in *Leviathan* "[b]y liberty is understood, according to the proper signification of the word, the absence of external impediments." *See* Hobbes ([1651], 104).

[25] Sheldon & Kasser (2001).

[26] Ryan et al. ([1996], 9).

[27] Sheldon et al. ([2004a], 219). Noting that "[I]n no culture did self-concordance correlate negatively with [flourishing] as a cultural relativist perspective might predict based on the assumption that self-possessed individuals do not 'fit' within collectivist societies."

[28] *Id.*

[29] Sheldon & Houser-Marko (2001).

[30] "Interest" is distinguished from "curiosity." Curiosity is associated with uncertainty and unpleasant affect. Interest, on the other hand, pertains to pleasurable affect that is associated with an activity. *See* Reeve ([1996], 170).

[31] For a discussion of core values see Loehr & Schwartz ([2003], 137–47).

[32] Ryan, et al. (1996).

[33] From William Butler Yeats' poem *The Second Coming,* available at: http://www.potw.org/archive/potw351.html

[34] Strictly speaking, the pleasure of contemplative activity is not unique to human beings since God experiences this supreme pleasure. However, amongst embodied creatures, the pleasure of contemplative activity is unique to human beings (*NE* 1178b).

[35] Granted, one could argue that the activity of smoking can arise from "peer pressure" or "external standards of coolness". For less contentious examples of habitual behaviors that emerge from an individual's voluntary actions, consult the chart on page 115.

[36] *See* Ryan & Deci (2000) for a discussion on the internalization and integration of external norms.

[37] *Id.*

[38] As Ryan and Deci point out, integration comes in degrees. Full integration occurs when external regulations are "fully assimilated into the self, which means they have been evaluated and brought into congruence with one's other values and needs." See Ryan & Deci ([2000], 73). So when Paul

understands the place of the norm "one ought to do one's homework" within his hierarchy of internalized norms, the norm is fully integrated.

Chapter 5

1 For a useful overview, *see*: http://en.wikipedia.org/wiki/Executive_functions
2 Eslinger (2003). *Available at*: http://www.cerebromente.org.br/n17/mente/brain-development.htm
3 Smith & Jonides (1999).
4 Woodward, et al. (2002).
5 Damasio ([1995], 8).
6 Squire, et al. ([2002], 1172).
7 Garon, et al. (2008).
8 *See Brain Briefings: The Adolescent Brain*, available at: http://www.sfn.org/index.aspx?pagename=brainBriefings_Adolescent_brain
9 *See* Miyake, et al. (2000).
10 Berk ([2003], 296).
11 Miyake, *supra* n. 9.
12 Bodrova & Leong ([2007], 68).
13 Berk ([2001], 75–106).
14 Berk ([1992], 21–22).
15 Berk, *supra* n. 13 at 81–82.
16 Berk ([2006a], 81–86).
17 Berk, *supra* n. 13 at 116–117.
18 *Id.*
19 Diamond et al. (2007). For an excellent National Pubic Radio story in which Dr. Diamond discusses the role of play in learning, *see* http://www.npr.org/templates/story/story.php?storyId=19212514
20 Berk, *supra* n. 13 at 116–117.
21 Berk (2001a). *Available at*: http://teachpsych.org/resources/e-books/faces/text/Ch10.htm
22 Metcalfe & Mischel (1999).
23 For research on the idea that will-power is a muscle see the excellent blog on this topic by Dr. Timothy Pychyl *available at*: http://www.psychologytoday.com/blog/dont-delay/200902/self-regulation-failure-part-2-willpower-is-muscle. Dr. Pychyl also has an excellent podcast on this topic *available at*: http://iprocrastinate.libsyn.com/
24 Wood, et al. (2005).
25 *Id.*
26 *Id.*
27 Martin (2008).
28 James ([1899], 42). See note 3, Chapter 3 (above) for information about accessing this text online.
29 *Id.* at 43.
30 Wood, *supra* n. 24.
31 Mischel & Ayduk (2002).
32 *Id.* at 114–116.

[33] *Id.*

[34] Nasar ([1998], 351).

[35] Epictetus (circa 101 AD), *Discourses*, see especially Book I, Chapter 27 and Book II, Chapter 16. *Available at*: http://classics.mit.edu/Epictetus/discourses.html

[36] Cicero (circa 45 BC), *Tusculan Disputations*, Book III, Chapter 33. *Available at*: http://www.gutenberg.org/etext/14988

[37] Peterson & Seligman (1984).

[38] *Id.*

[39] Anderson ([1983], 1145–6).

[40] Gollwitzer ([1999], 494).

[41] *Id.*

[42] Frankl ([1959], 66).

[43] Mischel & Ayduk, *supra* n. 31 at 114.

[44] Gollwitzer, *supra* n. 40 at 493–4.

[45] The marathoner adage is mentioned by Captain Chesley Sullenberger III, the U. S. Airways Pilot, who, after losing both engines (because Canadian geese flew into them), miraculously landed his jet on the Hudson River with no causalities. Captain Sullenberger mentions the marathoner adage in the context of climbing Mount Whitney with his wife, Lorrie. *See* Sullenberger ([2009], 176). To learn more about Captain Sullenberger, *see* http://en.wikipedia.org/wiki/Chesley_Sullenberger

[46] Latham & Locke (2007).

[47] *Id.*

[48] Snyder et al. ([2002a], 257–260). *See also* Snyder (2002b).

[49] Maddux (2002).

[50] *See* Reeve ([1996], 104–114) for discussion and references.

[51] Masten ([2001], 228).

[52] *Id.* at 234.

[53] For a description of the inertial guidance system (also referred to as an "inertial navigation system"), *see* http://en.wikipedia.org/wiki/Inertial_guidance. I am indebted to Jeffrey Wigand helping me to understand the nature of these systems.

[54] *See* Jim Loehr's discussion of the power of these questions at Loehr ([2007], 43–65).

[55] Ryan & Deci (2000).

[56] Research demonstrates that feeling understood and appreciated and talking about meaningful issues satisfy the relatedness need. *See* Reis, et al., (2000). Aristotle's notion of friendship is arguably the satisfaction of the relatedness need *par excellence*. True friends love each other for their own sakes (*NE* 1166a), and engage in discussion and thought (*NE* 1170b).

Chapter 6

[1] Duff ([2003], 214–5).

if re-framed as "study topic X" for 90 minutes. Specific goals are shown to both induce motivation and increase performance.[47]

Hope and self-efficacy

The framing ability is just one of the components of the motivational component of the executive function. Within positive psychology, the concepts of *hope* and *self-efficacy* refer to the motivational beliefs that induce somatic engagement. *Hope* is defined as the individual's perceived capability to derive pathways to desired goals, *and* the ability to motivate oneself via agency thinking to use those pathways.[48] *Self-efficacy* beliefs are encompassed by the first part of this definition. These beliefs consist in "people's beliefs in their capabilities to produce desired effects by their own actions."[49] If one daydreams about her self-concordant goal of becoming a philosophy professor, but does not believe that she could ever realize such a goal, she is unlikely to undertake and endure the work done to realize this goal. Voluminous research exists demonstrating that these beliefs are learned and that motivational deficits result when they are absent from one's belief set.[50]

Resilience

Another ingredient of the motivational component of the executive function is *resilience*. As mentioned above, practical reason operates within a dynamic domain characterized by uncertainty and variability. Within the shifting landscape of human affairs, tragedy and misfortune occur: the senseless death of a loved one, a terrorist attack, an economic crisis, to name but a few. Nevertheless, individuals have been able to bounce back from these difficult experiences. This is because such individuals possess *resilience*.

Resilience refers to a personal attribute which enables one to press on and persevere despite tragic circumstances. Anne Masten, one of the leading researchers on resilience theory, defines it as "a class of phenomena characterized by good outcomes in spite of serious threats to adaptation or development."[51] Note that resilience is an ability different from the ability to overcome stimulus control. We are

[2] Brian Little's important work on Personal Projects Analysis acknowledges the importance of ecology. *See* Little, et al. ([2007], 17–21).

[3] *Lawrence v. Texas.*, 539 U.S. 558 (2003). *Available at*: http://caselaw.lp.findlaw.com/scripts/getcase.pl?court=US&vol=000&invol=02-102#opinion1

[4] http://www.swsg.org

[5] *See* Markus & Nurius (1986) and Markus & Ruvolo (1989) for work on the connection between the possible self concept and motivation.

[6] Tough (2006). For more information about KIPP schools visit http://www.kipp.org/ *and* http://en.wikipedia.org/wiki/Knowledge_Is_Power_Program

[7] *Id.*

[8] McCown, et al. ([1998], 116–19).

[9] *See supra* n. 19, Chapter 5, for information about a National Public Radio broadcast that focused on the *Tools* program. Visit the website for the *Tools* program at: http://www.mscd.edu/extendedcampus/toolsofthemind/

[10] Bodrova & Leong ([2006], 150–1).

[11] *Id.*

[12] *See* Chapter 5, pp. 124–125.

[13] Gray ([2002], 10).

[14] *Id.*

[15] *Id.*

[16] *Id.* at 11.

[17] *Id.* at 10.

[18] *Id.* at 12.

[19] *Id.* at 13.

[20] *See,* e.g., Cohen (1998) and Trent (1998).

[21] One approach to life mapping involves a seven-step process that involves cutting out pictures from magazines. Such an approach is doubtless "light" on critical analysis. *See* http://www.oprah.com/article/spirit/knowyourself/ss_know_passion_01. A more rigorous mapping approach can be found in the excellent book by Loehr & Schwartz (2003).

[22] Abrams & Keren ([2007], 321).

[23] Captain Chesley Sullenburger is also an example of an individual who exemplifes self-concordance. In his book, *The Highest Duty,* Captain Sullenburger explains that the activity of flying expresses a deep interest and allows him to express one of his core values of helping people. *See* Sullenburger ([2009], 4–6, 19, and 149). *See supra* n.45, Chapter 5, for a brief discussion of Captain Sullenberger.

[24] http://www.teachforamerica.org/

[25] Kopp ([2001], 4).

[26] *Id.* at 191.

Bibliography

Abrams, K. & Keren, H. (2007), 'Law in the cultivation of hope', *California Law Review*, 95, 319–81.

Ackrill, J. L. (1980), 'Aristotle on *Eudaimonia*', in A. O. Rorty (ed.), *Essays on Aristotle's Ethics.* Berkeley, CA: University of California Press, pp. 15–33.

Anderson, C. (1983), 'Motivational and performance deficits in interpersonal settings: The effect of attributional style', *Journal of Personality and Social Psychology*, 45(5), 1136–47.

Annas, J. (2005), 'Virtue ethics', in D. Copp (ed.), *The Oxford Handbook of Ethical Theory*, New York: Oxford University Press, pp. 515–36.

Anscombe, G. E. M. (1958), 'Modern moral philosophy', *Philosophy*, 33(124), 1–19.

Anton, J. & Preus, A. (eds) (1983), *Essays in Ancient Greek Philosophy,* Albany, NY: State University of New York Press.

Baumeister, R.F. & Leary, M.R. (1995), 'The need to belong: Desire for interpersonal attachments as a fundamental human motivation', *Psychological Bulletin*, 117(3), 497–529.

Berk, L. (1992), 'Children's private speech: An overview of theory and status of research', in R. M. Diaz, & L. E. Berk (eds), *Private Speech: From Social Interaction to Self-Regulation*, Hillsdale, NJ: Lawrence Erlbaum Associates, Inc., pp. 17–54.

—(2001), *Awakening Children's Minds,* New York: Oxford.

—(2001a), 'Trends in Human Development' in Halonen, J. S., & Davis, S. F. (eds), The many faces of psychological research in the 21st century. *Available at*: http://teachpsych.org/resources/e-books/faces/script/Ch10.html. Retrieved November 17, 2009 from the e-book version of the text, *available at*: http://teachpsych.org/resources/e-books/faces/text/index.htm.

—(2003), *Child Development,* Boston, MA: Allyn & Bacon.

—(2006), *Development through the Lifespan,* Boston, MA: Allyn & Bacon.

Berk, L. et al. (2006), 'Make-Believe Play: Wellspring for Development of Self-Regulation' in R. Singer, & M. Golinkoff (eds), *Play=learning: How Play Motivates and Enhances Children's Cognitive and Social-Emotional Growth,* New York: Oxford University Press, pp. 74–100.

Bodrova, E. & Leong, D. (2006), *Tools of the Mind: The Vygotskian Approach to Early Childhood Education,* Columbus, OH: Merrill/Prentice Hall.

—(2007), 'Learning and development of preschool children from the Vygotskian perspective', in A. Kozulin (ed.), *Vygotsky's Educational Theory in Cultural Context,* Cambridge, UK: Cambridge University Press, pp. 156–76.

Burnyeat, M. (1980), 'Aristotle on learning to be good', in A. O. Rorty (ed.), *Essays on Aristotle's Ethics,* Berkeley, CA: University of California Press, pp. 69–92.

Cicero, (circa 45 BC/2002), *On the Emotions: Tusculan Disputations 3 and 4* (Margaret Graver, trans.), Chicago, IL: University of Chicago Press.

Cohen, B. (1998), *Life Mapping: A Unique Approach to Finding Your Vision and Reaching Your Potential,* New York: Harper Perennial.

Cooper, J. (1975), *Reason and the Human Good in Aristotle,* Cambridge, MA: Harvard University Press.

—(1987), 'Contemplation and happiness: A reconsideration', *Synthese,* 72, 187–216.

Csikszentmihalyi, M. (1990), *Flow: The Psychology of Optimal Experience,* New York: Harper Collins.

—(2000), *Beyond Boredom and Anxiety: Experiencing Flow in Work and Play,* San Francisco, CA: Josey Bass.

Damasio, A. (1995), *Descartes' Error: Emotion, Reason, and the Human Brain,* New York: Harper Perennial.

Davidson, R. J., et al. (1990), 'Approach/withdrawal and cerebral asymmetry: Emotional expression and brain physiology I,' *Journal of Personality and Social Psychology,* 58, 330–341.

Davidson, R. J. (1995), 'Cerebral Asymmetry, Emotion and Affective Style,' in R. J. Davidson & K. Hugdahl (eds), *Brain Asymmetry,* Cambridge, MA: MIT Press, pp. 361–387.

Deci, E. L. & Ryan, R. M. (2001), 'On happiness and human potentials: A review of research on hedonic and eudaimonic well-being', *Annual Review of Psychology,* 52, 141–66.

Deci, E.L. & Ryan, R.M. (eds) (2002), *Handbook of Self-Determination Research,* Rochester, NY: University of Rochester Press.

—(eds) (2002a), '*Overview of Self-Determination Theory: An Organismic Dialectical Perspective*', in E. Deci & Ryan, R. M. (eds), *Handbook of Self-Determination Research,* Rochester, NY: University of Rochester Press, pp. 3–33.

Denckla, M. B. (2007), 'Executive function: Binding together the definitions of attention-deficit/hyperactivity disorder and learning disabilities', in L. Meltzer (ed.), *Executive Function in Education,* New York: Guilford, pp. 5–18.

Devereux, D.T. (1981), 'Aristotle on the essence of happiness', in D. O'Meara, (ed.), *Studies in Aristotle,* Washington, D.C.: Catholic University Press of America, pp. 247–60.

Diamond, A. et al. (2007), 'Preschool program improves cognitive control', *Science,* 317, 1387–88.

Diener, E., et al. (1998), 'Subjective well-being is essential to well-being', *Psychological Inquiry,* 9, 33–7.

Duff, R. A. (2003), 'The limits of virtue jurisprudence', *Metaphilosophy,* 34, 214–24.

Epictetus (circa 101 AD/1916), *The Discourses and Manual* (P. E. Matheson, trans.), New York: Oxford University Press.

Eslinger, P. (2003), 'Brain development and learning', *Brain & Mind Magazine,* 17. *Available at*: http://www.cerebromente.org.br/n17/mente/brain-development.htm. Last accessed August 25, 2009.

Everson, S. (1997), *Aristotle on Perception,* New York: Oxford Clarendon Press.

Farrelly, C. & Solum, L. (2008), *Virtue Jurisprudence,* New York: Palgrave Macmillan.

Foot, P. (2002), *Virtues and Vices and Other Essays in Moral Philosophy*, New York: Oxford Clarendon Press.

Frankl, V. (1959/2006), *Man's Search for Meaning* (Ilse Lasch, trans.), Boston, MA: Beacon Press.

Gable, S. L. & Haidt, J. (2005), 'What (and why) is positive psychology?', *Review of General Psychology*, 9, 103–10.

Garon, N. et al. (2008), 'Executive function in preschoolers: A review using an integrative framework', *Psychological Bulletin*, 134(1), 31–60.

Goleman, D. (1997), *Emotional Intelligence*, New York: Bantam.

—(2003), *Destructive Emotions*, New York: Bantam.

Gollwitzer, P. M. (1999), 'Implementation intentions: Strong effects of simple plans', *American Psychologist*, 54, 493–503.

Gray, R. M. (2002), 'The Brooklyn Program: Innovative Approaches to Substance Abuse Treatment', 66-DEC Fed, 9–16.

Hala, S. (1997), *The Development of Social Cognition*, Psychology Press.

Halonen, J. S., & Davis, S. F. (eds) (2001), 'The many faces of psychological research in the 21st century', Retrieved November 17, 2009 from the Society for the Teaching of Psychology Web site: http://teachpsych.org/resources/e-books/faces/script/index.html

Hardie, W. F. R. (1965), 'The final good in Aristotle's ethics', *Philosophy*, XL, 277–95.

—(1980), *Aristotle's Ethical Theory*, 2nd edn. New York, Oxford: Clarendon Press.

Hobbes, T. (1651/1958), *Leviathan*, H. Schneider (ed.), Indianapolis, IN: Bobbs-Merrill.

Hume, D. (1748/1993), *An Enquiry Concerning Human Understanding*, E. Steinberg (ed.), Indianapolis, IN: Hackett.

Irwin, T. H. (1980), 'The metaphysical and psychological basis of Aristotle's ethics', in *Essays on Aristotle's Ethics*, A. O. Rorty (ed.), Berkeley, CA: University of California Press, pp. 35–53.

—(1988), *Aristotle's First Principles*. New York: Oxford Clarendon Press.

—(1991), 'The structure of Aristotelian happiness', *Ethics*, 101, 382–91.

James, W. (1899/2008), *Talks to Teachers on Psychology, and to Students on Some of Life's Ideals*, Sioux Falls, SD: NuVision.

Kandel, E. R. (2006), *In Search of Memory*. New York: W. W. Norton & Company.

Kenny, A. (1978), *The Aristotelian Ethics*. Oxford, UK: Clarendon Press.

Keyt, D. (1983), 'Intellectualism in Aristotle', in J. Anton & A. Preus (eds), *Essays in Ancient Greek Philosophy*, vol. 2., Albany: State University of New York Press, pp. 364–87.

Kolak, D. & Martin, R. (1999), *The Experience of Philosophy*, Belmont, CA: Wadsworth.

Kopp, W. (2001), *One Day All Children*, New York: Public Affairs Books.

Korsgaard, C. (1986), 'Aristotle on Function and Virtue', *History of Philosophy Quarterly*, 3, 259–79.

Kraut, R. (1989), *Aristotle on the Human Good*, Princeton, NJ: Princeton University Press.

—(1995), 'Reply to Professor Roche', in M. Sim (ed.), *The Crossroads of Norm and Nature*, Lanham, MD: Rowman & Littlefield, pp. 138–48.

Latham, G. P. & Locke, E. A. (2007), 'New developments in and directions for goal setting research', *European Psychologist*, 12(4), 290–300.

Lawson-Tancred, H. (1986), *Aristotle's* De Anima (a translation), New York: Penguin Books.

Little, B. R., Salmela-Aro, K. & Phillips, S. D. (eds) (2007), *Personal Project Pursuit: Goals, Action and Human Flourishing*, Mahwah, NJ: Lawrence Erlbaum Associates.

Loehr, J. (2007), *The Power of Story*, New York: Free Press.

Loehr, J. & Schwartz, T. (2003), *The Power of Full Engagement*, New York: Free Press.

McCown, K. S. et al. (1998), *Self-Science: The Emotional Intelligence Curriculum*, San Mateo, CA: Six Seconds.

Maddux, J. E. (2002), 'Self-efficacy: The power of believing you can', in Snyder, C. R. & Lopez, S. (eds), *Handbook of Positive Psychology*, New York: Oxford University Press, pp. 277–87.

MacIntyre, A. (1984), *After Virtue*, South Bend, IN: University of Notre Dame Press.

Martin, N. (2008), *Habit: The 95% of Behavior Marketers Ignore*, Upper Saddle River, NJ: Financial Times/Prentice Hall.

Markus, Hazel & Paula Nurius (1986), 'Possible selves,' *American Psychologist*, 41(9), 954–69.

Markus, H., & Ruvolo, A. (1989), 'Possible selves: Personalized representations of goals', in L. A. Pervin (ed.), *Goal concepts in personality and social psychology*, Hillsdale, NJ: Erlbaum, pp. 211–41.

Maslow, A. (1976), *Religions, Values and Peak Experiences*, New York: Penguin Books.

Masten, A. S. (2001), 'Ordinary magic: Resilience process in development', *American Psychologist*, 56, 227–348.

Meltzer, L. (2007), *Executive Function in Education*, New York: Guilford.

Metcalfe, J. & Mischel, W. (1999), 'A hot/cool system analysis of delay of gratification: Dynamics of willpower', *Psychological Review*, 106(1), 3–19.

Mischel, W. & Ayduk, O. (2002), 'Self-regulation in a cognitive-affective personality system: Attentional control in the service of the self', *Self and Identity*, 1, 113–20.

Miyake, A. et al. (2000), 'The unity and diversity of executive functions and their contributions to complex "frontal lobe" tasks: A latent variable analysis', *Cognitive Psychology*, 41, 49–100.

Modrak, D. K. (1976), '*Aisthesis* in the practical syllogism,' *Philosophical Studies*, 30, 379–91.

Nagel, T. (1980), 'Aristotle on *Eudaimonia*', in Rorty, A. O. (ed.), *Essays on Aristotle's Ethics*, Berkeley, CA: University of California Press, pp. 7–14.

Nasar, S. (1998), *A Beautiful Mind*, New York: Simon and Schuster.

O'Meara, D. (ed.), *Studies in Aristotle*, Washington, D.C.: Catholic University Press of America.

Peterson, C. & Seligman, M. (1984), 'Causal explanations as a risk factor for depression: Theory and evidence', *Psychological Review*, 91(3), 347–74.

Peterson, C. & Seligman, M. P. (2004), *Character Strengths and Virtues,* New York: Oxford University Press.

Purinton, J. (1998), 'Aristotle's definition of happiness', in *Oxford Studies in Ancient Philosophy,* 16, 259–378.

Reeve, C. D. C. (1995), *Practices of Reason,* Oxford, UK: Clarendon Press.

Reeve, J. (1996), *Motivating Others: Nurturing Inner Motivational Resources,* Boston, MA: Allyn & Bacon.

Reis, H.T. et al. (2000), 'Daily well-being: The role of autonomy, competence, and relatedness', *Personality and Social Psychology Bulletin,* 26, 419–35.

Roche, T. (1988), '*Ergon* and *Eudaimonia* in *Nicomachean Ethics* I: Reconsidering the Intellectualist Interpretation', *Journal of the History of Philosophy,* 26(2), 175–94.

—(1995), 'The ultimate end of action: A critique of Richard Kraut's *Aristotle on the Human Good*', in M. Sim (ed.), *The Crossroads of Norm and Nature,* Lanham, MD: Rowman & Littlefield, pp. 115–38.

Rorty, A. O. (ed.) (1980), *Essays on Aristotle's Ethics,* Berkeley, CA: University of California Press.

Ryan, R. M. & Deci, E. L. (2000), 'Self-determination theory and the facilitation of intrinsic motivation, social development, and well-being', *American Psychologist,* 55, 68–78.

—(2001), 'On happiness and human potentials: A review of research on hedonic and eudaimonic well-being', *Annual Review of Psychology,* 52, 141–66.

Ryan, et al. (1996), 'All goals are not created equal: An organismic perspective on the nature of goals and their regulation', in P. M. Gollwitzer & J. A. Bargh (eds), *The Psychology of Action: Linking Cognition and Motivation to Behavior,* New York: Guilford, pp. 7–26.

Ryff, C. & Keyes, C. (1995), 'The structure of psychological well-being revisited', *Journal of Personality and Social Psychology,* 69, 719–27.

Ryff, C. D. (1995), 'Psychological well-being in adult life', *Current Directions in Psychological Science,* 4(4), 99–104.

Sartre, J. P. (1957/1994), 'Existentialism', in *Existentialism and Human Emotions,* New York: Citadel Press.

Sheldon, K. et al. (2004), 'The independent effects of goal contents and motives on well-being: It's both what you pursue and why you pursue it', *Social Psychology Bulletin,* 30(4), 475–86.

Sheldon, K. et al. (2004a) 'Self-concordance and subjective well-being in four cultures', *Journal of Cross-Cultural Psychology,* 35(2), 209–23.

Sheldon, K. M. & Elliot, A. J. (1999), 'Goal striving, need-satisfaction, and longitudinal well-being: The Self-Concordance Model', *Journal of Personality and Social Psychology,* 76, 482–97.

Sheldon, K. & Houser-Marko, L. (2001). 'Self-concordance, goal-attainment, and the pursuit of happiness: Can there be an upward spiral?', *Journal of Personality and Social Psychology,* 80, 152–65.

Sheldon, K. & Kasser, T. (2001), 'Goals, congruence and positive well-being, new empirical support for humanistic theories', 41(1), *Journal of Humanistic Psychology,* 30–50.

Sheldon, K. & King, L. (2001), 'Why positive psychology is necessary', 56 *American Psychologist*, 216–217.

Shiner, R. (1979), '*Aisthesis, Nous,* and *Phronêsis* in the Practical Syllogism', *Philosophical Studies*, 36, 377–87.

Sim, M., (ed.) (1995), *The Crossroads of Norm and Nature*, Lanham, MD: Rowman & Littlefield.

Singer, D. G. et al. (eds) (2006), *Play=learning: How Play Motivates and Enhances Children's Cognitive and Social-Emotional Growth*, New York: Oxford University Press.

Smith, E. E. & Jonides, J. (1999), 'Storage and executive processes in the frontal lobes', *Science*, 283, 1657–61.

Snyder, C. R. & Lopez, S. J., (eds) (2002), *Handbook of Positive Psychology*, New York: Oxford University Press.

Snyder, C. R. et al. (2002a), 'Hope theory: A member of the positive psychology family', in C. R. Snyder & Shane Lopez (eds), *Handbook of Positive Psychology*, New York: Oxford University Press, pp. 257–76.

Snyder. C. R. (2002b), 'Hope theory: Rainbows of the mind', *Psychological Inquiry*, 13, 249–75.

Squire, L. R. & Kandel, E. R. (1999), *Memory: From Mind to Molecules*, New York: Henry Holt.

Squire, L. R. et al. (2002), *Fundamental Neuroscience*, New York: Academic Press.

Sullenberger, C. (2009), *Highest Duty: My Search for What Really Matters*, New York: Harper Collins.

Tough, P. (2006), '*What it Takes to Make a Student*', N.Y. Times Magazine, November 26, 9. *Available at*: http://www.nytimes.com/2006/11/26/magazine/26tough.html

Trent, J. (1998), *Life Mapping*, Colorado Springs, CO: Waterbrook Press.

White, R. (1959), 'Motivation reconsidered: The concept of competence', *Psychological Review*, 66(5), 297–333.

Whiting, J. (1986), 'Human nature and intellectualism in Aristotle', *Archiv fur Geschichte der Philosophie*, 68, 70–95.

—(1988), 'Aristotle's function argument: A defense', *Ancient Philosophy*, 8, 33–48.

Wilkes, K. (1980), 'The good man and the good for man in *Aristotle's Ethics*', in A. O. Rorty (ed.), *Essays on Aristotle's Ethics*, Berkeley, CA: University of California Press, pp. 15–33.

Wood, W. et al. (2002), 'Habits in everyday life: Thought, emotion and action', *Journal of Personality and Social Psychology*, 83(6), 1281–97.

—(2005), 'Changing circumstances, disrupting habits', *Journal of Personality and Social Psychology*, 88(6), 918–33.

Woodward, T. et al. (2002), 'Task switching deficits associated with Parkinson's disease reflect depleted attentional resources', *Neuropsychologia*, 40(12), 1948–55.

Index of Names

Annas, Julia 94–5
Anscombe, Elizabeth 93
Anthony, Susan B. 144, 145

Beckham, David 115
Berk, Laura 125
Bodrova, Elena 158

Cady Stanton, Elizabeth vi, 145
Cicero 135
Clinton, Hillary Rodham 101, 102, 114
Cooper, John 13
Cricket, Jiminy 29
Csikszentmihalyi, Mihalyi 96–7

Damasio, Antonio 121

Feinberg, Michael 158
Field, Justice Stephen 92
Foot, Phillipa 93
Frankl, Viktor 139

Gabriel, Peter 25
Gage, Phineas 121
Gates, Archie 62
Gibbons, Peter 1
Goleman, Daniel 92
Gollwitzer, Peter 136

Hagen, Walter 142
Hardie, W. F. R. 13
Hoffman, Philip-Seymour vi
Homer 101, 104
Hume, David 19
Hyde, Lindsay 157

Irwin, Terence 8

James, William 62–4, 128–9
Jones, Bobby 142

Kennedy, Justice Anthony 156
Kolak, Daniel 169, n.4
Kopp, Wendy 162–5
Kraut, Richard 15

Leong, Deborah 158
Levin, David 158

MacIntyre, Alasdair 93
Maslow, Abraham 96, 98
Masten, Anne 143
Michelangelo 73
Mill, John Stuart vi
Murray, W. H. 57

Nash, John 133–5

Porter, Cole 57
Pychyl, Timothy 125, n.23

Rogers, Carl 96

Sartre, Jean Paul 7–8, 41–3, 91, 95
Sullenberger, Chesley 172, n.45, 173, n.23

Tolstoy, Leo 39

Wigand, Jeffrey xiii, 172, n.53.

Index of Subjects

action 6–7, 26, 27, 40, 47–9, 50, 51–64, 66, 67–9, 72, 75, 85–7, 88, 93–4, 98, 102–9, 111, 117, 148–9, 151–2, 163–4, 165 *see also* causality *and* motivational system
  causality and 27–8, 52–7, 105
  of children and/or lower animals 51–2, 60–1, 67, 106, 109, 110, 113, 123–4, 124–5, 133, 158
  directed by practical reason 19, 22–3, 27, 30, 32, 50–2, 61, 64–6, 69, 80, 120–2, 150
  extrinsically motivated *see* motivation
  goal-directed *see* teleological/goal oriented activity: of the human being
  habitual *see* habit(s)
  intrinsically motivated *see* motivation
  rational (as contrasted with mere voluntary) 50–2, 60, 61, 68, 109, 110–12, 121
  of the student of the *NE see* Students of the NE
  voluntary 51–2, 107–12, 113, 117, 121, 124–5
action arousing/dampening attributes 132–3, 138–9, 146
Aesop's Fables 124
affect/affective part of soul xi, 47–8, 51–71, 86, 92, 96, 97, 104, 119, 120, 128, 146, 151 *see also* affective virtue, pleasure *and* proper pleasure
  and calling *see* calling
  causal power of 26, 53–6, 58, 59
  commanding nature of 47, 54–6, 61, 67–8, 69, 72, 86, 92, 106, 149–51
  correction/training of *see* motivational system: correction/transformation of
  default state of *see* motivational system: default orientation of
  and desire *see* causality *and* motivational system
  effect on belief 82–3
  effect on memory 81–2
  effect on perception 83
  incorrectness of *see* calling, apparent
  pleasurable *see* pleasure/pleasurable activity
  and positive psychology 97–8
  and its relationship to action *see* causality *and* motivational system
  and target of nature 59–60
affective training *see* motivational system: correction/transformation of
affective virtue 51, 58, 60–70, 75, 80–3, 84, 85, 87, 109, 116, 126–7, 129, 139, 151, 162
  development of *see* motivational system: correction/transformation of
akrasia 83
analogical reasoning 78
  in the *Function Argument* 5–6, 7–8, 13, 18, 38, 39–40, 41–3
analytic component of executive function *see* executive function
animals 35, 37, 44, 51–2, 59, 61, 73, 94, 109, 110

Anna Karenina 129
apparent good/calling *see* calling
affect: incorrectness of
Aristotle's view of the state 69, 153–5, 156, 161–2
attention 54, 120, 122, 125, 129, 131–5, 138–9, 140, 142, 144–6, 151, 159 *see also* private speech: steering function of
Attention Deficit Hyperactivity Disorder (ADHD) 120, 125
attributional style *see* explanatory style
authenticity 98, 104, 112–17, 159
autonomy xi, xii, 91–2, 95, 96, 98, 99, 100–1, 103, 104, 107–12, 113–18, 151, 152, 153, 155, 156, 161, 162 *see also* needs *and* self-concordance
automaticity *see* habit(s)

A Beautiful Mind 133
Bill & Melinda Gates Foundation 158
biology 94, 103
bodily pleasures *see* pleasure
boulêsis see wish
biological development *see* production: natural
Brooklyn Program 158, 159–60

calling
- and affective content 57–60, 69, 85, 106, 117, 149 *see also* affect/affective part of soul: commanding nature of
- apparent 41–3, 47, 57–60, 64, 69, 83, 106, 149, 151
- as a species of wish 28, 30–2
- of the craftsman *qua* craftsman 28, 29–32, 39–40, 41–4, 52, 136
- of the human *qua* human 31, 39–48, 54–5, 56–69, 84, 88–9, 91–2, 94, 99, 106, 148 149, 164
- of the student of the *NE* 48, 85, 88, 116–17
- versus target of biological development 33–8, 45, 59–60, 94

causality *see also* production
- Aristotle's theory of 19–23, 26
- efficient causality 20, 21, 26, 27–8, 30, 32, 35, 43, 49, 52–3, 55
  - and desire/affect 26, 27–8, 32, 49, 53, 55
- explanatory style *see* explanatory style
- final causality 18, 20–4, 26, 28, 30–2, 33–4, 36, 37, 39, 40–3, 45, 48–9, 52–6, 59, 70, 105
  - degrees of 22, 23, 28, 37, 57
  - and desire/affect 26, 30–1, 32, 52, 53–6, 105 *see also* calling
  - objective dimension 18, 33–4, 45 *see also* target
  - subjective/desiderative dimension 18–19, 28–31, 32, 33–4, 45 *see also* calling
- formal cause 20, 21, 36
- Hume's theory of 19
- material cause 20, 21
- perceived locus of *see* perceived locus of causality

Charleston 129
children 38, 51, 60–1, 67, 69, 73, 84, 106, 109, 113, 123–5, 133, 158, 160, 164
choice 27–8, 30, 32, 39, 40, 42, 52, 55, 85, 127, 136 *see also* implementation intentions
commanding nature of affect *see* affect
community *see* political community
competence *see* needs
contemplation/contemplative activity xi, xii, 2, 3, 4, 12–17, 39, 45, 46–8, 50, 58–9, 60, 61, 70–3, 79–80, 83, 84, 87–8, 90–1, 100, 109–11, 118, 148, 165
- development of xi, xii, 45, 46, 47–8, 60–1, 70–85, 88, 116–18, 148

continence/incontinence 49, 65, 127
"cool" attributes *see* action arousing/dampening attributes
correction of the default affective state *see* motivational system: correction/transformation of

De Anima (*DA*) 26, 36, 48, 50, 74, 78
Declaration of Independence 91
default affective state *see* affect
default interpretive system 150–1
default orientation of motivational system *see* motivational system
deliberate desire *see* choice
deliberation *see* practical deliberation
Descartes' Error 121
desire/desiderative capacity of the human soul 23, 26–32, 33–4, 38–9, 40–6, 47, 48, 49, 50–1, 52–8, 59–61, 65–7, 80, 83, 84, 85, 87, 88, 93–4, 99–101, 105, 127, 136, 149, 150 *see also* affect, choice, wish *and* calling
desire to know 99–100 *see also* motivation: towards rational activity
developmentalism xi, xii, 5, 14, 17, 162
dialectic 70, 76–9, 81
DNA 4, 36, 48 *see also* genes/genetic activity

ecology 144, 165
  legal/political 152, 153–7, 162
  moral 153–4
  normative 116–17, 153–4
education 155–6, 157, 161, 162, 163–4
educational inequality 163–4
efficient causality *see* causality
Emotional Intelligence 92
emotions *see* affect
emperia see universals: primitive
endoxa 78–9
epikeia see equity
epistêmê see knowledge
equity (*epikeia*) 81
ergon 5–11, 43–4, 83, 118, 148
ethical virtue/ethically virtuous activity xii, 3–4, 12–17, 39, 45, 46, 47–8, 70, 71, 72, 79–80, 83, 84–8 *see also phronêsis*
eudaimonia see flourishing
excessive affect/passion 67, 81–3
executive function 119, 120–1, 145, 148, 161 *see also* practical reason
  intellectual component of 121–2, 124, 145, 146, 147, 148, 154, 158, 161, 162, 164 *see also* strategic planning
  motivational component of 120, 141–6, 147, 148, 154, 158, 159–60, 161, 162, 164
  navigational component of 120, 122–41, 146, 147, 148, 154, 157, 158, 161, 162, 164 *see also* self-regulation
experience (*emperia*) *see* universals: primitive
experience sampling method (ESM) 96–7
explanatory style 135, 138, 139, 146
external habit triggers *see* habit(s)
extrinsic motivation *see* motivation

feedback 122
final causality *see* causality
first principles 37, 76, 78–9
flourishing (*eudaimonia*) xi–xii, 1–16, 18–19, 20, 22, 31, 33–4, 38–9, 40–3, 45, 46, 47–8, 54, 56, 59, 61, 69, 84, 85, 88, 90–100, 104, 106, 107, 109, 110, 111, 116–17, 118, 128, 147–50, 152, 153–7, 160, 161, 162
  eudaimonic model of 98
  hedonic model of 97–8
flow 96, 97
form
  and man-made production 24–6, 31–2, 36
  and natural production/soul 5, 35–8, 48–9
formal causality *see* causality
framing 142, 144, 146
free exercise of religion 95
free play 124–5
friendship 82, 100, 108

function argument xii, 2–11, 13, 14, 17, 18, 20, 38–40, 42–5, 59
fundamental needs *see* needs

Generation of Animals (*GA*) 17, 38
genes/genetic activity 36, 75 *see also* DNA
goal attainment *see* teleological/goal oriented activity *and* self-concordance
goal intention (contrast with implementation intention) 136
goal oriented activity *see* teleological/goal oriented activity
good/goods of action 7–8, 28, 40–3, 50, 57, 104–5 *see also* calling

habit(s) 62–4, 66–7, 100, 107, 108, 111, 113–17, 113–30, 126–9, 130–5, 137–9, 140–1, 146, 149–51, 153–6, 154, 158
  authentic 112–16 *see also* internalization *and* integration
  development of virtuous *see* motivational system: correction/transformation of
  integrated 114, 115
  normative 113–17
  primitive 113, 115
  and self-regulation *see* self-regulation: masterful
  as a source of pleasure 64, 66–7
  of the students of the *NE* *see* students of the NE: habits of
  triggers 128–9, 130–3, 136–41, 144, 146, 151
    external 133, 137, 139, 140, 141, 146, 151
    internal 133, 134–5, 138–9, 140, 141, 146, 151
habituation *see* motivational system: correction/transformation of
Harvard Law School 57
hope 29, 144, 146, 157, 158, 159
"hot" attributes *see* action arousing/dampening attributes
human nature xi, 4, 8, 42, 68, 91, 94, 99, 100, 150
hypothetical necessity 18–19, 21–2, 23–5, 28, 32, 33–5, 37, 140, 164 *see also* causality *and* production

idion 5, 9–11, 17, 37–9, 42, 43, 44, 59
Iliad 101, 104
implementation intentions 136–41, 144–6 *see also* choice
impulse control *see* self-restraint
inclusivist (interpretation of the *NE*) 2, 14–17, 39, 45, 46, 48, 87
induction 76–9
inertial guidance system 146
inhibition *see* self-restraint
integrated habits *see* habit(s): integrated
integration (of norms) 114–17
intellectual/analytic component of executive function *see* executive function
intellectualist (interpretation of the *NE*) xi, 2, 5, 14–15, 16, 17, 39, 41, 45, 46, 70, 84, 85, 87–8
internal habit triggers *see* habit(s): triggers
internalization (of norms) 86, 114–17
interpretive system 150–1 *see also* explanatory style
intrinsic motivation *see* motivation
intrinsically desirable goods 104–5 *see also* motivation: intrinsic

KIPP schools *see* Knowledge is Power Program
knowledge (*epistêmê*) *see also* perception
  of the craftsman *see* form and man-made production
  as a precondition of *sophia* 45, 46–8, 72–9
Knowledge is Power Program (KIPP) 158

law 69, 93, 116, 151, 153–7, 162 *see also* ecology: legal
Lawrence vs. Texas 156
Legend of Bagger Vance 142
liberalism xi, xii, 95, 100–1, 152, 153–7, 162
liberty *see* autonomy
life coaching 160
life mapping 161
Living History 101–2, 112, 114

man-made production *see* production
marshmallow test 133
masterful motivation *see* motivation
masterful self-regulation *see* self-regulation
material causality *see* causality
memory 63, 64, 73–6, 78, 82, 120, 122
  neurobiology of 74–6
meta-cognition 133, 134, 138
Metaphysics (*Met.*) 70, 78
Michigan Bar Exam 55–6
mindfulness 66, 68, 124, 129–31, 133–4, 137–9, 144, 145–6, 151, 158
molecular biology 4
moral development xi, 48, 65, 87, 112 *see also* motivational system: correction/transformation of
motivation *see also* motivational system
  as a component of the executive function *see* executive function: motivational component of
  extrinsic 102, 112–13, 150 *see also* perceived locus of causality
  intrinsic 47, 68, 72, 98, 101–7, 112, 117, 119, 142, 149, 150–1, 163 *see also* perceived locus of causality
  masterful 144–6, 163
  towards rational activity 59–60, 66–8, 71–2, 79–80, 83, 86–8, 99–100, 106–7, 111, 149 *see also* desire to know
motivational component of executive function *see* executive function: motivational component of
motivational system 19, 33, 34, 36–7, 87–8, 99, 104–5, 106, 118, 128, 148, 149–51, 155, 157, 164
  correction/transformation of 60–70, 75, 106–7, 113–16, 118, 126, 127, 128, 149–52, 164 *see also* masterful motivation *and* masterful self-regulation
    public component of 61–3, 126
    private component of 64–6, 126
  of the craftsman *qua* craftsman 19, 23–32, 39, 41–3, 52, 136
  default orientation of 61–6, 69, 86, 106, 128, 149–51, 164
  desiderative components of 26–32
  of the human *qua* human 34, 40–5, 47–8, 52–61, 68–70, 84, 87–8, 99, 106, 118, 128, 148, 149–51, 155, 157, 164
  intellectual components of 23–6
  as a precondition to contemplation *see* contemplation: development
  of the students of the *NE* 85–7, 116–17

natural production *see* production
navigational component of executive function *see* executive function
needs
  for autonomy 98, 99, 100–4, 151
  for competence 98, 99, 100
  for rational activity *see* desire to know
  for relatedeness 98, 99, 100, 148
  versus desires 99–100
nervous system 62–3, 74, 128
neural network 74–5
neurobiology 74–6

neurons 74–5
neuroscience 74, 119, 120, 121
neurotransmitters 74
Nike 141
normative ecology *see* ecology
normative habits *see* habit(s)
norms 124, 153 *see also* Internalization *and* Integration *and* Ecology, Normative
nous 38, 59, 60, 72
Nueva School 158
nutritive capacity of the soul *see* soul

Office Space 1
On Memory 74, 82
One Day All Children 163
"other regarding" virtues *see* virtues

pain 47, 52, 53–4, 56, 57, 58, 60, 63–5, 66, 68, 69, 71, 82, 99, 135, 164 *see also* affect
Parkinson's disease 120
particular statements (as opposed to universal statements) 72–3, 77, 78
particulars (objects of opinion/ perception) 72, 77
Parts of Animals 17
perceived locus of causality 101–2 *see also* motivation, intrinsic; motivation, extrinsic *and* self-perception
perception 9, 70, 72–6, 78, 79–84 *see also* knowledge
phronêsis xii, 3–4, 12, 39, 44, 46, 47, 48, 49, 60–1, 66–73, 80–4, 87, 127, 145, 162 *see also* ethical virtue/ethically virtuous activity
phronimos 66, 67, 68, 70, 80–1, 83, 85, 165
Physics (*Phys.*) 17, 20, 78
play *see* free play
pleasure/pleasurable activity 27, 29, 30–1, 32, 41, 42, 47, 52, 53, 57–61, 63–70, 71–2, 80, 82, 86–8, 90, 92, 97–8, 99–100, 101–2, 103–17, 118, 127, 149–52, 155, 164, 165 *see also* affect, affective virtue, desire, *and* proper pleasure
  bodily 42, 56, 58, 60, 61, 63, 67, 68, 83, 85, 106, 116, 127, 150
  and calling 30–2, 42–3, 52, 58–60, 69, 106, 116–17, 148, 149
  causal power of *see* affect: causal Power of
  incorrectness of *see* calling, apparent
  proper *see* proper pleasure
  and wish 29–30, 32
political community 95, 96, 153–6, 161
Politics 99, 116–17
polity 152, 154, 155, 156, 157, 161, 162
positive psychology xi, xii, 92, 96–8, 143
possible selves 155, 158
postmodernism 91
practical reason 3, 23–7, 28, 30, 32, 42, 50–2, 60–1, 80–1, 109, 110, 119, 120, 121, 126, 143, 145, 150 *see also* executive function *and* strategic planning
practical wisdom *see phronêsis*
preconditions of contemplative excellence *see* contemplation: development
prefrontal cortex 120, 121
prohairesis see choice
primitive universals *see* universals
Princeton University 163
private speech 123–8 *see also* self-regulation
  Aristotle on 64–6, 126–8
  development of 124–5
  and implementation intentions 137–9, 144–6
  steering function of 125–8, 131–5, 137–9, 144–6, 151, 157 *see also* attention, masterful motivation, *and* self-regulation

production *see also* causality *and* hypothetical necessity
  man-made 18–32, 33–4, 36, 37, 38, 39, 42, 43, 44, 45
  natural (biological development) xi, xii, 4–5, 14, 17, 18, 19, 33–8, 43–5, 48, 59, 94, 121 *see also* target
proper pleasure 59–61, 66–70, 71, 79–80, 86–7, 92, 98, 105–11, 116, 118, 149–52, 155, 164
  awakening of 66–70, 79–80, 86–7, 106–7, 118, 149–52, 155
  developmental role of 70–2, 87–8
psychology xi, 92, 94, 96, 119, 120 *see also* positive psychology

rational activity *see* soul: various rational activities of
rational part(s) of the soul *see* soul
rational pleasure *see* proper pleasure
reason *see* soul: various rational activities of
relatedness *see* needs
resilience 143–4, 146, 157
responsibility 111–12
Rhetoric (*Rhet.*) 82
right to free speech 95
right to privacy 95, 156
Rubin vase-profile illusion 131–2
rule following 51, 52, 60, 65–6, 68–9, 75, 110, 124, 126–7 *see also* self-restraint *and* habit(s): normative

secularism 156
self-concordance xii, 90, 102–4, 107, 114, 115, 118, 119, 120, 125, 135, 140–1, 143, 146–52, 153–62, 155, 157, 161–2, 163–5
  development of 118, 119–47, 149–52, 153–62
  example of 163–5
self-concordant activity *see* Self-Concordance
self-concordant goals *see* Self-Concordance
self-determination *see* autonomy
self-determination theory (SDT) xii, 92–3, 98–104, 107, 111, 114, 117–18, 148–9, 161
self-efficacy 143–4, 146, 157, 158, 159, 160
self-knowledge 141, 147, 148, 151
  Development of 147
self-perception
  of autonomous activity 103–11, 112–17
  of extrinsically motivated activity *see* perceived locus of causality
  of intrinsically motivated activity *see* perceived locus of causality
  of virtuous activity 107–9, 112–17
self-regulation 120, 123–41, 151, 152, 157, 158 *see also* executive function: navigational component of *and* private speech
  development of 124–8, 158–9
  masterful 127–9, 130–41, 144, 146, 158, 162
self restraint 51–2, 60, 65, 67, 120, 123–6, 127, 145, 146, 147, 158 *see also* rule following, *and* self-regulation
Self-Science curriculum 158
self-talk *see* private speech
set-shifting 122
somatic engagement 141–2
sophia 3, 4, 12, 72, 84, 87, 88
sophos 45, 48, 70, 72, 84, 87
soul (Aristotle's theory of) 35–8
  affective part of *see* affect
  and biological development: *see* production, natural
  as efficient cause of body 35
  as form/final cause of the body 35–8, 48
  nutritive capacity of 9, 49, 59
  rational parts of 10, 11, 50, 66, 67, 68, 70, 121 *see also* practical reason *and sophia*

as a set of capacities 46, 48–9
various rational activities of 1–2, 3–4, 11–12, 13–16, 24, 39, 49–50, 51, 64–6, 70–9, 80–1, 86–8, 109–11, 119, 121, 145 *see also* contemplation, dialectic, equity, ethical virtue, induction, practical deliberation, strategic planning, *phronêsis, and sunêsis*
South Bronx 163
Steering function of private speech *see* private speech
stimulus control 126, 130, 131–41, 143
stoic school of philosophy 135
strategic planning 23–6, 30, 31, 32, 50–2, 65, 67, 80, 121, 145, 146, 161, 164 *see also* practical deliberation
Strong Women, Strong Girls 157
students of the *NE* xi, 48, 85–8, 111–14, 115–17, 153
habits of 86–6, 111–17
subjective well being 96, 97
sunêsis 81
synapse 75

Talks to Teachers on Psychology (William James) 62
target (of biological development) 33–8, 43–5, 59–60, 84, 94 *see also* calling *and* production: natural
Teach for America 163–4
teleios/teleiotatos 2–3, 10, 11, 12–13, 14, 15–16, 18, 39, 45
teleological/goal oriented activity *see also* causality: final
of the human being 23, 31, 33, 45, 66, 100–4, 117, 119–23, 133, 136–7, 145–7, 148, 155, 156, 162 *see also* calling *and* motivation: extrinsic *and* intrinsic
of nature *see* production: natural *and* target
theoria see contemplation
Three Kings 62
Tools of the Mind curriculum 158
transformative aim of the *NE* xi, 48, 85–9, 112, 117
triggers *see* habit(s)
Twitter 130

universal statements (as opposed to particular statements) 72–3, 76–9
universals 73–6, 77–9, 82
primitive (*emperia*/experience) 73–9, 81–2

virtue ethics *see* virtue theory
virtue jurisprudence 93
virtues
of character *see* affective virtue *and* ethical virtue
Judaeo-Christian conception of 148
modern (preconditions of self-concordance) xi, xii, 98, 118, 147–8, 152, 161
other regarding 2, 84–5, 87, 148
virtue theory xi–xii, 93–5, 98, 118, 148, 161
voluntary action *see* action

War and Peace, 39
will power 125, 126, 129, 131, 133, 151
wish (*boulêsis*) 29–32, 38–40, 42, 52, 57, 136 *see also* calling

LaVergne, TN USA
10 May 2010
182144LV00002B/6/P